Nitrogen Management in Crop Production

Editors

Mr. Pankaj K. Dubey and Prahlad Maurya

scitus
academics

Nitrogen Management in Crop Production

Edited by **Pankaj K. Dubey and Prahlad Maurya**

ISBN: 978-1-68117-024-4

Library of Congress Control Number: 2015931393

SCITUS Academics LLC,
www.scitusacademics.com
Box No. 4766, 616 Corporate Way,
Suite 2, Valley Cottage,
NY 10989

Notice

Printed in United States of America on Acid Free Paper ∞

Contents

Preface

Sustainable development has increased the demand for knowledge on fertilizer management in crop production as one of the main strategies for safeguarding food security. Among the necessary plant nutrients, nitrogen is mainly responsible for determining the yield and yield of legumes and cereals. Sound nitrogen management for crops makes great financial sense. Optimal fertilization of nitrogen is crucial to achieving grain yield and quality objectives and achieving maximum economic returns. Excess fertilization increases the risk of accommodation and lowers total income. Good nitrogen management also has good environmental significance. Excess fertilizer application of nitrogen increases the environmental loss of nitrogen, including emission leaching of nitrate and the groundwater of greenhouse gas nitrous oxide. The supply of nitrogen to a cereal crop comes from fertilizer, mineralization and manure. Mineralization is the release of plant nitrogen from organic soil material and crop residues as a result of microbial soil activity. The optimal fertilizer nitrogen rate varies from field to field and year to year due to variations in the demand for both crops and the supply of nitrogen to soil.

This book covers the key elements for the proper use of nitrogen in crop systems. This includes appropriate timing methods, effective source and application of nitrogen during crop growth cycles, the use of a sufficient application rate to avoid damage and lower costs, the use of nitrogen-efficient crop genotypes and the use of legumes that fix adequate amounts of nitrogen in the atmosphere.

Chapter 1

Quantification of Plant Stress using Remote Sensing Observations and Crop Models: The Case of Nitrogen Management

F Baret, V Houlès and M Guérif

INRA-CSE, Site Agroparc, F-84914 Avignon, France

ABSTRACT

Remote sensing techniques offer a unique solution for mapping stress and monitoring its time-course. This article reviews the main issues to be addressed for quantifying stress level from remote sensing observations, and to mitigate its impact on crop production by managing cultural practices. The case of nitrogen fertilization is used here as a paradigm. The derivation of canopy state variables such as the leaf area index (*LAI*) and chlorophyll content (C_{ab}) is first addressed. It is demonstrated that the inversion of radiative transfer models leads to useful estimates of these variables. However, because of the ill-posed nature of the inverse problem, better accuracy is achieved when using prior information on the distribution of the variables and when multiplying *LAI* by C_{ab} to get canopy level chlorophyll content. This variable, $LAI \times C_{ab}$ is well suited for quantifying canopy level nitrogen content. It is used for nitrogen stress evaluation by comparison with a reference unstressed situation which is, however, not easy to get in practice. The combination of remote sensing observations with crop models provides an elegant solution for stress quantification through assimilation approaches. It fuses several sources of information within our knowledge of the processes involved and accounts for the environmental budget which can be integrated when making decisions about cultural practices. Conclusions are drawn on the issues related to the retrieval of canopy state variables from remote sensing data, to the link between these observables and crop models, and to the assimilation approaches. Avenues for further research are finally discussed along with the required observation system.

INTRODUCTION

Stresses come from a range of factors that limit the potential growth of canopies. Evaluation of the stress level to which plants are subjected is therefore critical information required both for the quantification of consequences on production and for taking action for their mitigation. Many stresses are caused directly or indirectly by water shortages or insufficient satisfaction of canopy mineral requirements, particularly nitrogen. Fungi and pests may also induce stresses by reducing the green leaf area, the sap flow, or by altering plant metabolism. Therefore, most stresses show-up with similar symptoms: (i) leaf area index (*LAI*) reduction, due both to a decrease of leaf area production and to an increased senescence rate; (ii) chlorophyll content (C_{ab}) decrease due either to a limitation of the synthesis of chloroplasts, to their destruction, or to the remobilization of this important protein pool; (iii) canopy temperature increase due to lower transpiration fluxes. Stress will therefore be expressed by changes in a few canopy state variables such as *LAI*, chlorophyll content, or surface temperature. However, clear identification of the stress factors is not always possible without using additional information. The signal emitted or reflected by vegetation surfaces is governed by state variables describing canopy structure (leaf area and orientation, spatial arrangement, roughness) and on the optical, dielectric, or thermal characteristics of the vegetation elements. The physical quantity recorded by remote sensors depends on the spectral domain considered where different physical processes may be involved: (i) in visible, near and short-wave infrared domains, the physical quantity considered is reflectance, i.e. the fraction of radiation reflected by the surface; (ii) in thermal infrared, canopy is characterized by its brightness temperature corresponding to the radiation flux emitted; (iii) in active microwaves, the back-scattering coefficient is measured by radar systems; (iv) in passive microwaves, the canopy is characterized similarly to the thermal infrared domain by its brightness temperature where however, emissivity plays a key role.

Each spectral domain is therefore sensitive to specific canopy state variables (Table 1). Canopy structure variables are mainly accessible in the solar reflective domain, i.e. from the visible to the short-wave infrared. Microwaves are sensitive to water content (Paloscia *et al.*, 1999; Prévot *et al.*, 2003; Wigneron *et al.*, 2004). Chlorophyll and water content in the leaf may be estimated with a reasonable accuracy from the solar reflective domain (Combal *et al.*, 2002; Ustin *et al.*, 2004) because of strong and specific light absorption bands. The thermal infrared domain provides information on leaf surface temperature governed by the energy and water balance of the canopy (Olioso *et al.*, 2002). For stress detection and quantification, the reflective solar domain appears very pertinent, particularly regarding the capacity to estimate leaf area index (*LAI*) and chlorophyll content (C_{ab}). Both variables are widely used by agronomists and ecologists to diagnose possible stress affecting the canopies. This is the basis for applications to agricultural management techniques developed recently, such as precision farming for which the spatial heterogeneity within and between fields is explicitly accounted for through site-specific practices (Cook and Bramley, 1998; Moran *et al.*, 1997; Pierce and Nowak, 1999). Remote systems sample the radiance field reflected or emitted by canopies at spatial resolutions and frequencies suitable for such applications, i.e. from sub-metric to metric and deca-metric pixel size, allowing mapping the spatial heterogeneity and the associated stress levels. Sensors in the visible and near infrared spectral domains are available aboard a range of vectors, from satellite, airborne, balloons, tractors, down to hand-held devices. They sample the temporal dimension over a range of frequencies, although cloud occurrence is a serious limitation for observations in the solar domain from high altitude vectors.

Table 1: Canopy state variables accessible from remote sensing observations in several spectral domains

Biophysical variables	Visible and near infrared	Near and short wave infrared	Thermal infrared	Active μ– waves (radar)	Passive μ-waves
Canopy structure					
Leaf area (LA!)	+++	+++	+	++	+
Cover fraction	++++	++++	++	++	+
fAPAR (fraction of absorbed PAR)	++++	++	++	++	+
Leaf orientation	++	++	+	+	+
Dimensions/ height	+	+	+	+	+
Canopy water content		++		+++	++
Leaf					
Water content		+++		+	+
Temperature			++++		+
Chlorophyll content (C_{ab})	+++				

Deriving stress maps from the observed radiance field is a complex process (Guérif *et al.*, 2001). The radiance field should first be translated into canopy state variable maps such as *LAI* or chlorophyll content. Then the stress level quantification at each point of the map may be achieved by comparison with unstressed canopies placed in similar conditions. This assumes that unstressed areas are known and precisely located in the fields, and representative of all the field conditions apart from the targeted stress, which is obviously difficult to organize. Other references are thus required, and crop models may provide a good alternative: they allow accounting for a number of forcing variables such as climate, soil or cultural practices. They also allow quantifying the stress level and its impact on the final production. However, models are not perfect and require a large number of parameters and initial conditions not always very well-known and that may vary spatially. Combining radiometric information with crop models within an assimilation approach is very appealing for stress quantification. In addition, crop models may be used in a prognostic mode to evaluate several scenarios of cultural practices under variable climatic conditions and to select the optimal one that maximizes the economic output for the farmer, while preserving the environment.

This article reviews the several steps required to go from remote sensing observations to stress quantification and making optimal decisions in agriculture. The approach is illustrated by nitrogen stress which is one of the main drivers of yield for many species. Although the unit price of nitrogen fertilizer is generally low regarding the possible impact on yield and grain quality, excess nitrogen may strongly impact the environment through leaching in the water table, and may also lead to lodging and increased vulnerability to several diseases. It is therefore pertinent to develop methods allowing sustainable nitrogen fertilization management.

The problems associated with the derivation of canopy state variables from remote sensing observations are addressed first. Then several ways to link these canopy state variables to stress levels are described. Finally, the combination of remote sensing observations and crop models is investigated, with the application to stress quantification and the optimal control of cultural practices.

Estimation of canopy state variables from remote sensing observations

The spectral and directional distribution of top of canopy reflectance at a given time and location is governed by the canopy structure, optical properties of the elements including the soil background, as well as view and illumination geometrical configuration. Because of numerous variables influencing canopy radiometric response, it is not straightforward to extract specific canopy variables. Radiative transfer models simulate the reflectance field as a function of the whole set of influencing variables including *LAI* and C_{ab}. However, retrieving these variables of interest from measurements of top of canopy reflectance as sampled with a particular sensor, requires inverting radiative transfer models.

Figure 1 presents the general scheme of canopy variable retrieval from top of canopy (TOC) remote sensing observations. A radiative transfer model simulates in the forward direction the TOC reflectance field from the input variables. These were split into the variables of interest such as *LAI* andC_{ab} and the other characteristics (geometrical view and illumination configuration, other canopy variables such as leaf inclination, leaf water content, soil optical properties…). An inverse technique is then used to extract the variables of interest. It generally requires prior information on the distribution of the input variables to regularize the inversion process (Tarentola, 1987).

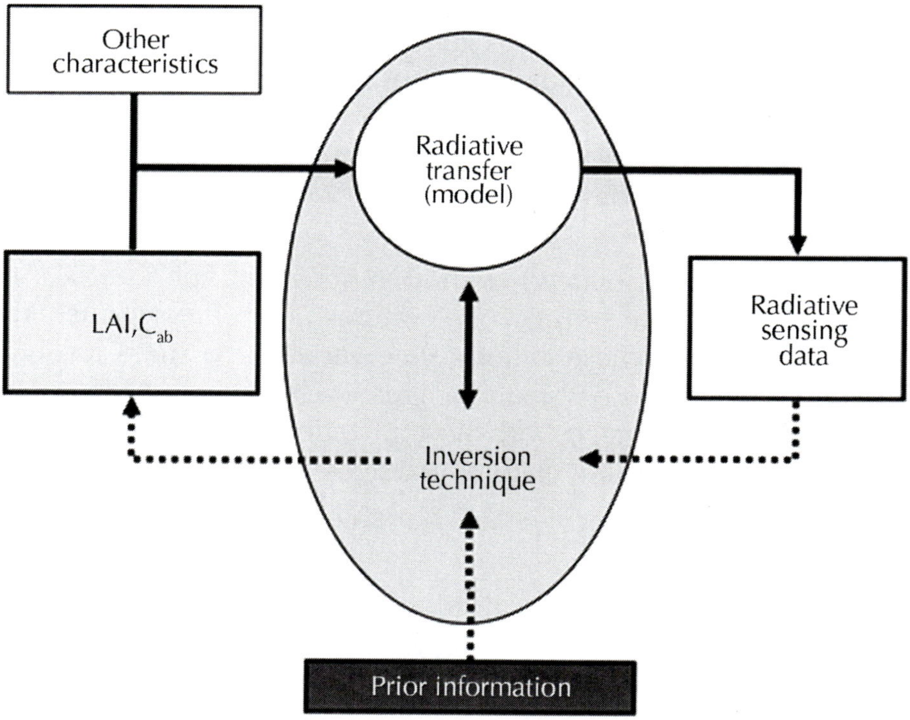

Figure 1: Scheme showing the inputs and outputs required in the forward and inverse problems to estimate canopy state variables (in this case *LAI* and C_{ab}).

As a matter of fact, several combinations of input variables may provide very similar reflectance simulations that match closely the actual remote sensing observations as illustrated by Figure 2. This corresponds to the generally ill-posed nature of the inverse problem in remote sensing (Combal *et al.*, 2002). The radiometric information is not sufficient to identify a unique solution: the inverse problem needs to be regularized by exploiting additional information such as prior knowledge on the statistical distribution of canopy radiative transfer model input variables. Closer inspection of the possible solutions (Figure 2) shows a strong negative correlation between the retrieved *LAI* and C_{ab} values. All the solutions have very similar values of canopy integrated chlorophyll content corresponding to the $LAI \times C_{ab}$ product between leaf level chlorophyll content (C_{ab}) and *LAI*. Note that canopy integrated chlorophyll content is a physically sound quantity since it represents the optical path in the canopy where absorption by chlorophyll governs the radiometric signal. Note also that canopy integrated chlorophyll content neglects chlorophyll contained in other elements than leaves such as stems.

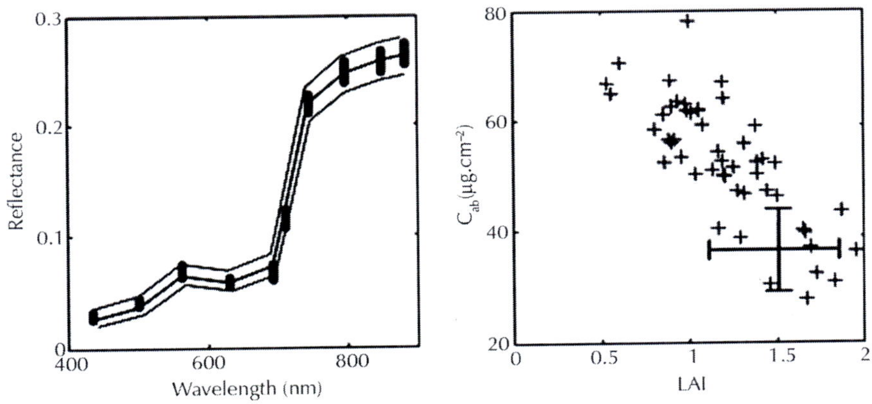

Figure 2: Actual reflectance measurements (left plot, solid lines, with ±standard deviation) and the corresponding closer simulations (dots) achieved with SAIL turbid medium radiative transfer model (Verhoef, 1984). On the right, the input *LAI* and C_{ab} (the '+' symbols) variables used to simulate the reflectance spectra shown on the left plot. The actual *LAI* and C_{ab} measurements (the large cross) are shown with their associated confidence intervals (corresponding to 1 standard deviation). Data acquired over a sugarbeet experiment conducted in 1990. More details are provided in Combal *et al.* (2001).

When prior information on canopy and soil input variables are exploited within the inversion process, improvement of the retrieval performances is observed (Figure 3). The accuracy and robustness of canopy characteristics estimation is even improved when using canopy level chlorophyll content ($LAI \times C_{ab}$) as compared to leaf level contents (C_{ab}). In the following, the interest in using canopy integrated chlorophyll content variable will be further demonstrated with reference to its role as a pertinent diagnostic variable.

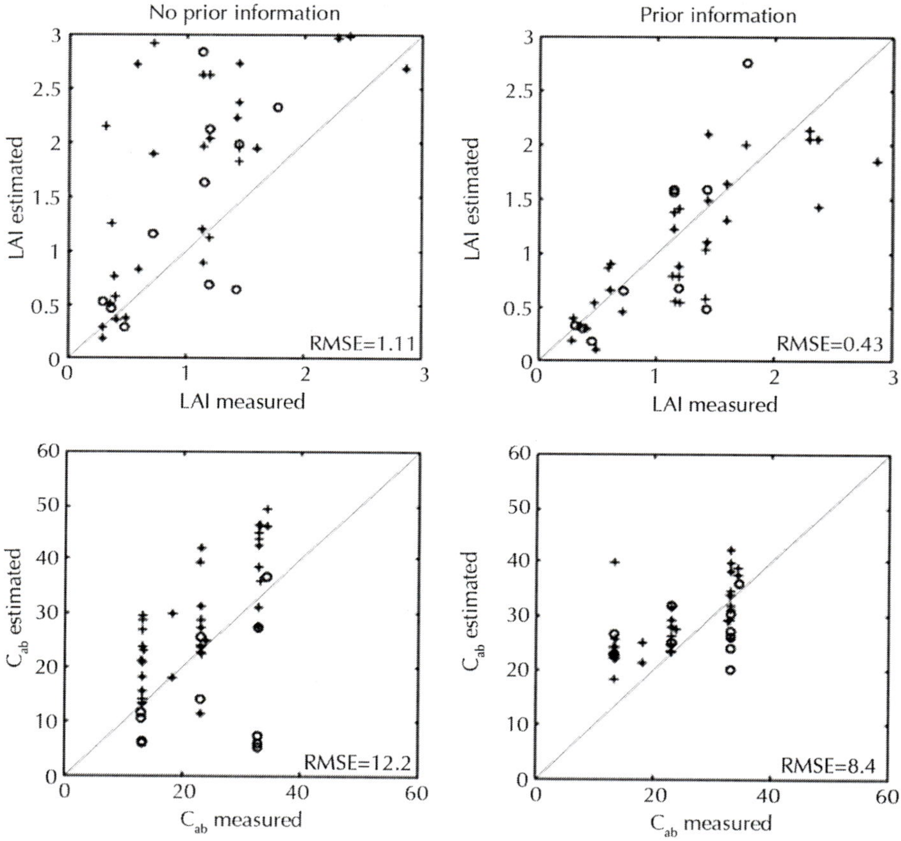

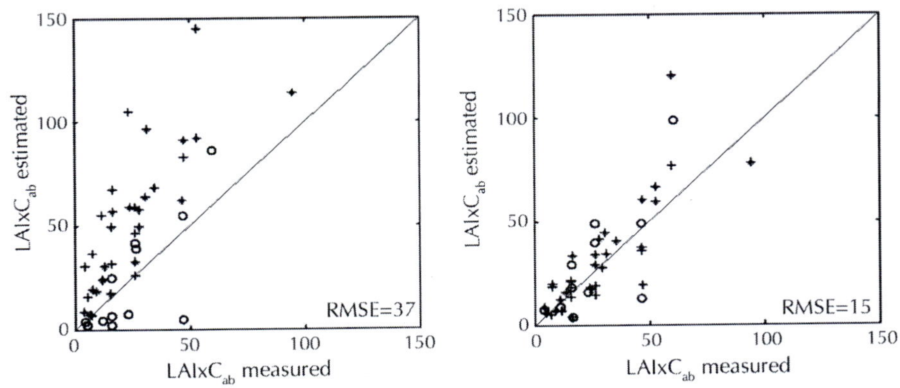

Figure 3: Comparison of retrieval performances for *LAI* (top),C_{ab} (middle), and $LAI \times C_{ab}$(bottom) over sugarbeet canopies when using no prior information (left) or with prior information. More details can be found in-Combal *et al.* (2001).

Linking remote sensing estimates of canopy variables to stress diagnostic variables: the case of nitrogen stress. The first study conducted to evaluate the use of remote sensing for nitrogen stress quantification were mainly based on empirical relationships with spectral indices that were sensitive to chlorophyll content (Penuelas *et al.*, 1994). Other authors (Bausch and Duke, 1996) use leaf chlorophyll content estimated with the SPAD instrument (Ercoli *et al.*, 1993) to evaluate canopy nitrogen status with reference to a non-stressed canopy. More recently, Vouillot *et al.* (1998) directly relate the nitrogen nutrition index as defined by Justes *et al.* (1994) and Lemaire *et al.* (1989) to the ratio between red and near infrared reflectance. However, as demonstrated previously, canopy radiometric response depends on many variables, and there is little chance that such a simple ratio between two bands shows robust relationships with some of the biophysical variables of the canopy as demonstrated by several other studies (Baret and Guyot, 1991). Nevertheless, the concept of optimal nitrogen concentration proves pertinent since it constitutes the reference case required to identify nitrogen stress and even to quantify it. Nitrogen nutrition index (*NNI*) is defined with regard to the optimal nitrogen concentration, N_R, achieving maximum biomass production (Figure 4). $N_R$values depend on the biomass amount according to the dilution theory (Lemaire and Gastal, 1997):

$$N_R = 5.35 \quad \text{if } W \leqslant 1 \text{ t ha}^{-1}$$
$$N_R = 5.35W^{-0.442} \quad \text{if } W \geqslant 1 \text{ t ha}^{-1}$$

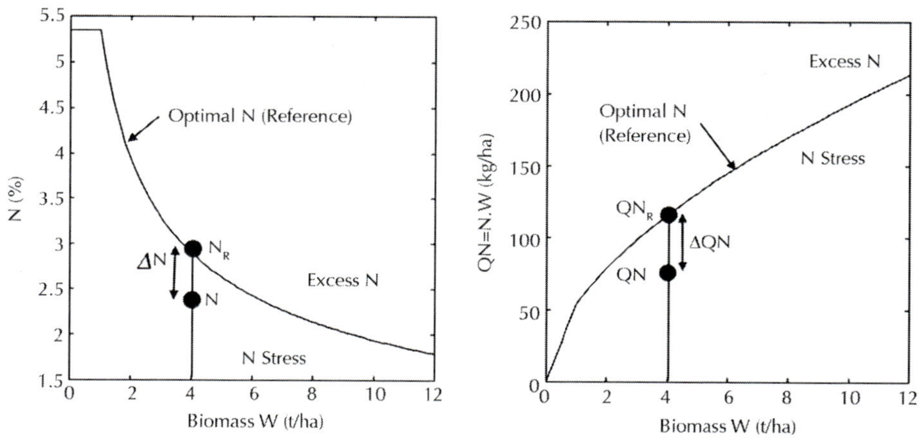

Figure 4: Critical curve corresponding to the optimal nitrogen concentration in the plant as a function of plant biomass (left plot). On the right, the optimal canopy nitrogen content is shown as a function of the biomass.

The nitrogen nutrition index is simply the ratio between the actual nitrogen concentration, N, to the optimal value, N_R: $NNI=N/N_R$. If $N > N_R$, then $NNI > 1.0$ and the canopy will not use efficiently the nitrogen amount which is in excess. Conversely, a canopy is assumed to be stressed if the nitrogen concentration is below the optimal value, i.e. $NNI<1.0$. The amount of nitrogen required by the canopy, assuming a nitrogen use efficiency of 1.0, will be $\Delta QN=W(\Delta N)$, where W is the actual canopy biomass and $\Delta N=N_R-N$. It would therefore be more useful to express the optimal nitrogen level in terms of nitrogen content (kg ha^{-1}) rather than in terms of concentration (%). Figure 4, right, illustrates the concept of optimal nitrogen content, QN_R, with $QN_R=WN_R$.

Agronomists have been focusing mostly on nitrogen content or concentration measurements to diagnose possible stresses. However, Baret and Fourty (1997) demonstrated that there were very few chances to retrieve canopy nitrogen directly or protein content from remote sensing observations, even using hyper-spectral systems. Nitrogen status could only be accessed through chlorophyll estimates which can actually be retrieved with reasonable accuracy as seen earlier in this paper. A dedicated experiment conducted in Laon in 2000 and 2001 over wheat crops subjected to a range of nitrogen stresses clearly demonstrated the interest of relating nitrogen to chlorophyll at canopy level rather than at leaf level (Houlès *et al.*, 2006). The relationships obtained at canopy level between $LAI{\times}C_{ab}$ and QN are more robust among years and development stages than the relations obtained between N and C_{ab} (Figure 5). However, these relationships depend slightly on development stages: after earing, plants show larger amounts of nitrogen for the same chlorophyll amount when a large part of nitrogen is concentrated in the grains (Figure 5). The younger stages also have more nitrogen for the same amount of chlorophyll and this might be due similarly to the dilution of nitrogen in the structural biomass with leaf development as confirmed by the strong relationship between leaf nitrogen and leaf chlorophyll content per unit soil area (Figure 6). This clearly shows the complexity of the processes associated with nitrogen and chlorophyll distribution in plants.

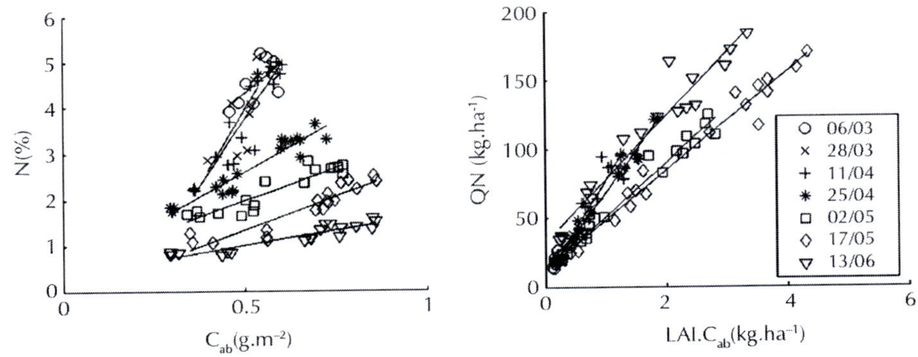

Figure 5: Relationships between C_{ab} and N (top), and betweenLAI × C_{ab} and QN (bottom) as observed over wheat experiments conducted at Laon (France) in 2001 (Houlès, 2004).

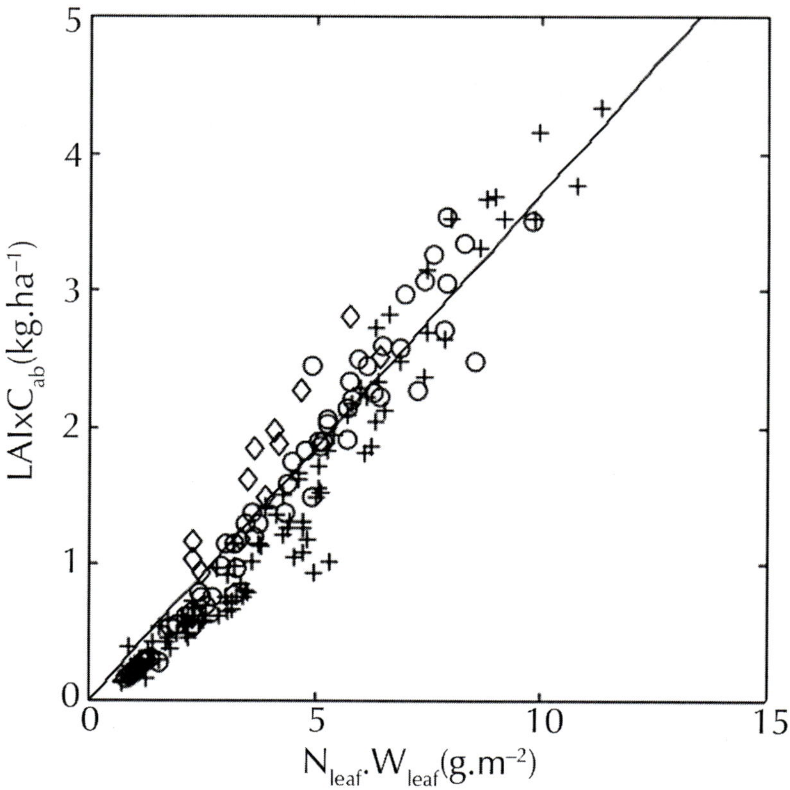

Figure 6: Relationships between leaf nitrogen content per unit soil area ($N_{leaf} \times W_{leaf}$) and chlorophyll content ($LAI \times C_{ab}$). Laon experiment conducted in 2000 and 2001 (light symbols) (Houlès, 2004) and ADAM experiment conducted in Fundulea (Romania) in 2001 (dark symbols) (Baret *et al.*, 2001). '+' and 'o' symbols correspond, respectively, to before and after earing stages.

Canopy integrated chlorophyll content therefore offers great potential for detecting and quantifying nitrogen stresses. The dilution curve is used here as a reference to evaluate a possible deficit in nitrogen availability when the actual biomass amount is known. This biomass can be derived from remote sensing measurements either through empirical relationships established during crop development with *LAI*, or by using simple light use efficiency models (Ruimy *et al.*, 1994) where the fraction of *PAR* absorbed by the canopy could be derived from remote sensing observations (Weisset *al.*, 2000). However, the stability of these empirical relationships, or that of the light use efficiency is questionable when plants are subjected to non-optimal conditions with regard to water balance or to unusual climatic conditions, or when considering new cultivars. Using crop models accounting for all these effects may thus constitute an elegant alternative solution for setting the reference case required for stress level evaluation. Crop models may also simulate nitrogen use efficiency that may vary widely depending on climatic and soil conditions. Remote sensing observations will then provide a way to correct the trajectory of crop models when all the initial conditions or soil and canopy parameters are not perfectly known.

Using Crop Models for Stress Management

Several authors have proposed using crop models as diagnostic tools (Acock and Pachepsky, 1997; Engel, 1997; Matthews and Blackmore, 1997; Bouma *et al.*, 1999; Kropff *et al.*, 2001; Batchelor *et al.*, 2002; Jones*et al.*, 2003). However, crop models require a relatively large number of input variables, initial conditions, and parameters. They may also suffer from structural deficiencies (Boote *et al.*, 1996). They therefore have difficulties in operational applications where only a fraction of these required variables and parameters are precisely known. This is particularly true when describing the spatial variability within fields, and crop models generally show poor performances (Hayes and Privette, 1998). However, remote sensing observations can be used to keep models on track as suggested by Jongschaap (2006). Such assimilation approaches are emerging recently due to the combination of advances in remote sensing data and interpretation methods, more realistic crop models, and large improvement in computation performance (Guérif and Duke, 1998;Nouvellon *et al.*, 2001; Bach and Mauser, 2003; Launay and Guérif, 2003;Prevot *et al.*, 2003). These models, once recalibrated using remote sensing observations could be used in prognostic mode to select the best strategies for crop management (Moore and Tyndale-Biscoe, 1999; Houlès*et al.*, 2004).

In the following, we will illustrate how to exploit a crop model to quantify nitrogen stresses and will propose an optimal strategy for fertilizer application in the context of precision agriculture (Houlès, 2004). The STICS model is used here to simulate canopy functioning as a function of climatic, soil, and cultural practices (Brisson *et al.*, 1998, 2002). The approach is divided into two steps. (i) Recalibration of the model inputs and parameters using remote sensing observations. This corresponds to the assimilation step. (ii) Selection of the optimal scenario by running in prognostic mode the recalibrated model. This corresponds to the control step.

Assimilation of Remote Sensing Observations

In the context of precision agriculture, the main issue consists in assessing the within-field variability of the main driving soil and crop variables and parameters. Therefore, a sensitivity analysis (Ruget *et al.*, 2002) was conducted to select the most important ones that need calibration. The other variables and parameters were assessed from expert knowledge. Cultural practices and climatic variables measured during the experiments were used as model inputs. Recalibration of crop models is an inverse problem which is difficult to use, similar to that of the canopy state variables retrieval from remote sensing. The GLUE Bayesian method was therefore used (Beven and Binsley, 1992; Makowski *et al.*, 2002), exploiting prior distribution of the variables and parameters to be estimated to get the optimal values for their later distribution. This method is summarized into the following four steps (for more details, seeHoulès *et al.*, 2004; Guérif *et al.*, 2006) (Figure 7). (i) A large number of parameter vectors (200 000) are randomly drawn within their prior distribution defined by expert knowledge and previous experiments. Note that at this stage, very little information is available on within-field spatial distribution of the parameters. (ii) The output variables corresponding to*LAI* and *QN* observations, derived previously from remote sensing observations, are simulated for each parameter vector using the STICS model. (iii) Observed *LAI* and *QN* values are then compared with the simulated values, and the likelihood of each parameter vector is computed. (iv) Finally, the later distribution of each parameter is estimated using the weights attributed to each parameter vector as derived from its corresponding likelihood.

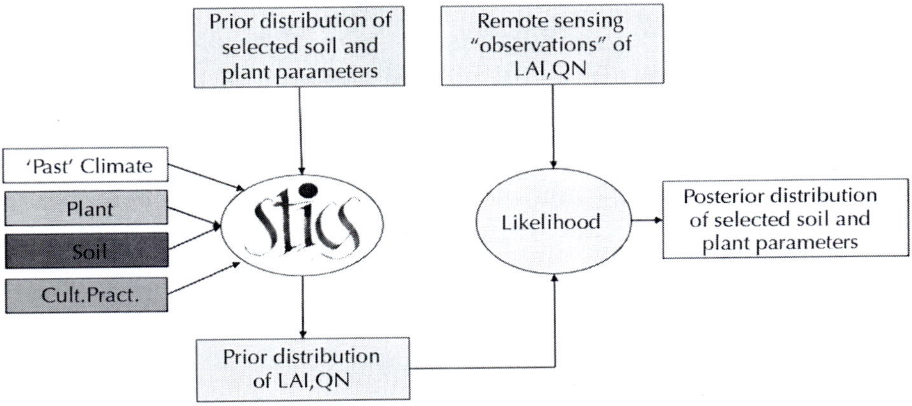

Figure 7: Scheme illustrating the assimilation achieved using the GLUE method.

The application of this technique is illustrated here over a wheat experiment (Houlès, 2004) where *LAI* and $LAI \times C_{ab}$ were estimated from airborne remote sensing measurements during four flights covering the growing season (Moulin *et al.*, 2003). *QN* values were then derived from remote estimates of $LAI \times C_{ab}$ based on empirical relationships. The technique was applied to all the pixels (20 m×20 m) of the field. When using only prior information on the unknown variables, results show no spatial structure of *LAI*, *QN*, and yield at each of the four dates (Figure 8, left), because prior distribution of the unknowns was only statistical, without incorporating the lack of knowledge on their spatial structure. Conversely, when remote sensing estimates of *LAI* and *QN* are assimilated within STICS, the spatial structure appears clearly (Figure 8, right). The agreement between measured and simulated *LAI* and *QN* values improves significantly, the points being roughly aligned along the 1:1 line. This result was expected for *LAI* and *QN* values, since the assimilation consists in tuning the unknown soil and cultivar parameters to get the best match between STICS simulated *LAI* and *QN* values, and those derived from remote sensing. The *RMSE* value for *LAI* decreases from 1.04 to 0.57 (Figure 8, top) after assimilation. This is not the case for *QN* (Figure 8, top) where an increase in the *RMSE* value (from 33.4 to 40.2) is observed after assimilation. This may indicate that either the observations, the formalism of the model or the prior information used are not correct. This is more relevant for *QN*, where the simulated values after assimilation are significantly smaller than remote sensing estimates of *QN* for the higher *QN* values corresponding to the two latest flights. Conversely, the two first flights (smaller *QN* values) show good consistency with remote sensing *QN* values. However, the decrease of *RMSE* values observed for yield (from 1.81 to 0.73, Figure 8, bottom) shows that the assimilation was efficient for this key crop model output. Inspection of the standard deviations (the vertical bars in Figure 8) shows also that the assimilation of remote sensing observations decreases the standard deviation in some cases (*QN*, first flights, yield, low values).

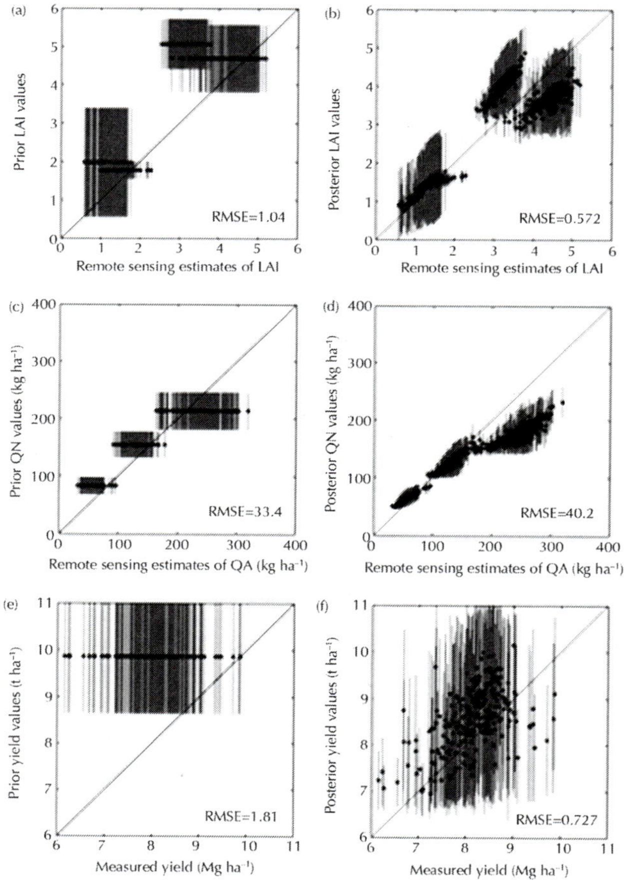

Figure 8: On the left, comparison between the measured values of *LAI*, crop nitrogen content (*QN*) and yield with their corresponding prior distribution. On the right, measured values of *LAI*, crop nitrogen content (*QN*) and yield with the corresponding posterior distribution obtained after assimilation of remote sensing estimates of *LAI* and *QN*. The vertical bars indicate the standard deviation.

Control of Cultural Practices

The control of cultural practices will be illustrated here for the third nitrogen application over the previous wheat experiment for which posterior distribution of the unknown parameters were adjusted as described above using the four remote sensing flights. To optimize nitrogen application over each part of the field, eight scenarios of nitrogen amount were compared (Figure 9). Because of the unknown weather conditions between the time of the decision and harvest, 30 climate scenarios were considered using climatic records from previous years. The simulations were then performed for all the combinations of climate and nitrogen scenarios for each of the 250 pixels of 20×20 m^2 covering the 10 ha of the field. This resulted in $8 \times 30 \times 250 = 60\,000$ simulations. Note that at this level, the later distribution of each unknown parameter was simply approximated by its median value. Although the whole distribution was available from the previous 'assimilation' step, using the median value reduces drastically the number of simulations and saves a very significant amount of computer time, at the expense of a slight degradation of the accuracy of the simulations.

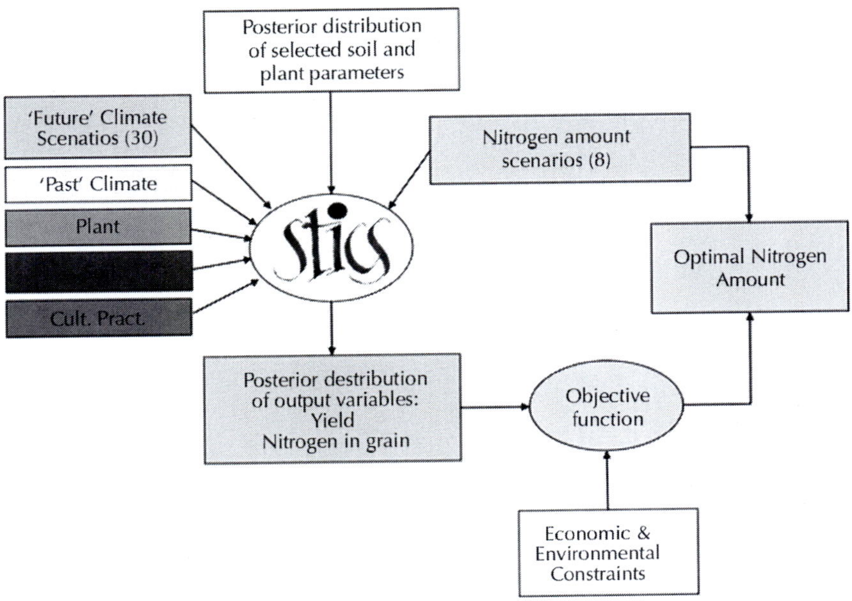

Figure 9: Scheme illustrating how to use of a crop model for the management of nitrogen applications.

The STICS model simulated the corresponding yield (*Y*) and grain protein content (*Ng*). These outputs were used to compute the corresponding gross margin, *Gm*: $Gm = Y \times Pg(Ng) - N \times P_N$ where *Pg*(*Ng*) is the price of a unit mass of grain as a function of the quality measured by the protein content. *N* is the amount of nitrogen applied, and P_N the unit price of nitrogen. Although there are currently no strong environmental policy regulations aimed at reducing nitrogen leaching in the soil by applying lower nitrogen amounts, crop models offer interesting potential for such evaluations. Houlès *et al.* (2004) showed that a simple budget computed as the difference between nitrogen inputs (applications) and outputs (nitrogen in the grain: $Y \times Ng$) can be used as an environmental constraint. Consequently, the objective function proposed consisted of discarding all nitrogen rates leading to a value of the budget exceeding a given threshold and then selecting the nitrogen scenario maximizing the gross margin. Note that the resulting optimal nitrogen amounts accounted for the climatic variability represented by 30 climate scenarios.

This process applied over all the pixels of the field provides the optimal nitrogen application map as illustrated by Figure 10. Note that no nitrogen application was required over about 20% of the field area, because canopy requirements are small enough to be covered by soil contribution. An homogeneous nitrogen application would lead to excess nitrogen left in the soil with some possible leaching into the water table. Conversely, applying a unique average nitrogen rate to the areas requiring the larger nitrogen amounts would reduce the yield and the farmer's income.

Figure 10: Optimal nitrogen application map (Nitrogen amounts in kg ha^{-1} for the third application) as derived from the combination of remote sensing observations and crop models.

CONCLUSIONS

Nitrogen is used here as a paradigm for stress management exploiting remote sensing observations. Most methods dedicated to nitrogen management are based on some measurements or estimates of canopy nitrogen content to decide how much nitrogen is to be applied at each location in the field. This canopy nitrogen content is compared to a reference level corresponding to an optimal nitrogen nutrition situation. The difference indicates the actual stress level and therefore the required amount of nitrogen to be applied. One of the main problems related to these techniques is the availability of such reference non-stressed situations, and the generally poorly known nitrogen use efficiency. Farmers may therefore set aside a small part of their field to apply important amounts of nitrogen to ensure unstressed reference status. However, this field sample is not necessarily representative of the other field conditions: spatial heterogeneity may be due both to variation in the soil properties, and to canopy status in relation to changes in practices (sowing depth and density, variability of previous applications ...). In addition, the amount of nitrogen required by the canopy also depends on the potential yield and nitrogen use efficiency, which vary with soil characteristics and the climatic conditions that will prevail after the nitrogen application up to the final harvest.

Crop models offer a very promising way to estimate this reference level and its variability in the field. In addition, crop models are able to account for several scenarios of climate conditions prevailing between the nitrogen application considered and harvest, frequency analysis of these variables allowing a better decision to be derived. It can also account for nitrogen use efficiency. Running crop models in such a predictive mode is very appealing for managing cultural practices as illustrated in this paper. Because models may simulate production in quantity and quality, as well as environmental variables (stock of mineral nitrogen left in the soil, for example), the decision may account for both economic and environmental constraints. It may even be used to design and tune policies dedicated to the sustainability of agricultural practices and systems.

However, the success of crop models for decision making relies on their performances for yield and environmental budget simulations. Uncertainties in model outputs depend on errors in the structure of the model and in some of its parameters and variables characterizing the initial conditions. Many of these parameters and variables vary from place to place in the field, and for most of them, there is no direct way of measuring them. The integration of remote sensing observations within such crop models would reduce these uncertainties. Remote sensing observations in the visible and near infrared spectral domains allow canopy leaf area index and leaf chlorophyll content to be mapped. It is demonstrated here that the canopy chlorophyll content is more strongly related to the canopy nitrogen content. This provides the necessary link between remote sensing observations and canopy state variables used as indicators of nitrogen status. Fortunately, chlorophyll content is better estimated at the canopy than at the leaf levels. This is mainly explained by the possible compensations observed in the inversion process between leaf level chlorophyll content and leaf area index: several combinations of *LAI* and C_{ab} may lead to very similar spectral reflectance responses. However, the uncertainties associated with canopy nitrogen content are still significant, around 20–30 kg ha^{-1}, which is at the limit of the acceptable level. Sources of uncertainties come first from the remote sensing estimates of canopy chlorophyll content, both because of measurement uncertainties and of the lack of realism of the radiative transfer models used. In addition, the relationships between canopy and nitrogen chlorophyll contents may vary slightly depending on the situations encountered.

Further research should therefore be dedicated to the improvement of these estimates from a better description of canopy architecture in the radiative transfer models, including possible gradients in chlorophyll content, and a proper modelling of the relationships between chlorophyll and nitrogen at the canopy level. These research axes should converge towards the development of the coupling between crop models and dynamic description of canopy 3D architecture, itself coupled to a radiative transfer model. Once this whole modelling chain has been developed, remote sensing observations could be directly assimilated in terms of radiance quantities in such process models. The only remaining limiting factor will be the availability of timely observations. Technological advances now allow systems to be designed with a cluster of satellites making global and frequent observations with a spatial resolution between 5–20 m which is required for most agriculture applications.

REFERENCES

1. Acock B, Pachepsky Y. Precision agriculture 1997. Oxford: Bios Scientific Publishers Ltd; 1997. Holes in precision farming: mechanistic crop models; p. 397-403.

2. Bach H, Mauser W. Methods and examples for remote sensing data assimilation in land surface process modeling. IEEE Transactions on Geoscience and Remote Sensing 2003;41:1629-1637.

3. Baret F, Fourty T. Radiometric estimates of nitrogen status of leaves and canopies. In: Lemaire G, editor. Diagnosis of nitrogen status in crops. Heidelberg, Germany: Springer-Verlag; 1997. p. 201-227.

4. Baret F, Guyot G. Potentials and limits of vegetation indices for *LAI* and APAR assessment. Remote Sensing of the Environment 1991;35:161-173.

5. Baret F, Vintila R, Lazar C, et al. The ADAM database and its potential to investigate high temporal sampling acquisition at high spatial resolution for the monitoring of agricultural crops. Romanian Agricultural Research 2001;16:69-80.

6. Batchelor WD, Basso B, Paz JO. Examples of strategies to analyse spatial and temporal yield variability using crop models. European Journal of Agronomy2002;18:141-158.

7. Bausch WC, Duke HR. Remote sensing of plant nitrogen status in corn.Transactions of the ASAE 1996;39:1869-1875.

8. Beven KJ, Binsley AM. The future of distributed models: model calibration and uncertainty predictions. Hydrological Processes 1992;6:279-298.

9. Boote KJ, Jones JW, Pickering NB. Potential uses and limitations of crop models.Agronomy Journal 1996;88:704-716.

10. Bouma J, Stoorvogel J, Van Alphen BJ, Booltink HWG. Pedology, precision agriculture, and the changing paradigm of agricultural research. Soil Science Society of America Journal 1999;63:1763-1768.

11. Brisson N, Mary B, Ripoche D, et al. STICS: a generic model for the simulation of crops and their water and nitrogen balances. I. theory and parameterization applied to wheat and corn. Agronomie 1998;18:311-346.

12. Brisson N, Ruget F, Gate P, et al. STICS: a generic model for simulating crops and their water and nitrogen balances. II. Model validation for wheat and maize.Agronomie 2002;22:69-92.

13. Combal B, Baret F, Weiss M. Improving canopy variables estimation from remote sensing data by exploiting ancillary

information. Case study on sugar beet canopies.Agronomie 2001;22:2-15.

14. Combal B, Baret F, Weiss M, Trubuil A, Macé D, Pragnère A, Myneni R, KnyazikhinY, Wang L. Retrieval of canopy biophysical variables from bi-directional reflectance data. Using prior information to solve the ill-posed inverse problem. Remote Sensing of Environment 2002;84:1-15.
15. Cook SE, Bramley GV. Precision agriculture: opportunities, benefits and pitfalls of site-specific crop management in Australia. Australian Journal of Experimental Agriculture 1998;38:753-763.
16. Engel T. Precision agriculture 1997. First European Conference on Precision Agriculture, *Vol. I.* Spatial variability in soil and crop. Oxford, UK: Bios Scientific Publishers Ltd; 1997. Use of nitrogen simulation models for site-specific nitrogen fertilization; p. 361-369.
17. Ercoli L, Mariotti M, Masoni A, Massantini F. Relationship between nitrogen and chlorophyll content and spectral properties in maize leaves. European Journal of Agronomy 1993;2:113-117.
18. Guérif M, Baret F, Moulin S, Bégué A. Télédétection, hétérogénéité parcel*lai*re et gestion spatialisée des interventions techniques. In: Malézieux E, Trébuil G, JaegerM, editors. Modélisation des agrosystèmes et aide à la décision. Montpellier:Repères; 2001. p. 303-326.
19. Guérif M, Duke C. Calibration of SUCROS emergence and early growth module for sugarbeet through optical remote sensing data assimilation. European Journal of Agronomy 1998;9:127-136.
20. Guérif M, Houles V, Makovski D, Lauvernet C. Data assimilation and parameter estimation for precision agriculture using the crop model STICS. In: Wallach D,Makovski D, Jones JW, editors. Working with dynamic crop models. Evaluation, analysis, parameterization and applications. Amsterdam: Elsevier; 2006. p. 391-398.
21. Hayes JC, Privette I. Potential benefits of precision farming. Seventh International Conference on Computers in Agriculture 1998:70-78.
22. HoulèsV. Mise au point d'un outil de modulation intra-parcellaire de la fertilisation azotée du blé d'hiver basé sur la télédétection et

un modèle de culture.Saint-Mandé (France): Université de Marne la Vallée; 2004.

23. Houlès V, Guérif M, Mary B, Gate P, Machet JM, Moulin S. Elaboration d'un indicateur de nutrition azotée du blé basé sur l'indice foliaire et la teneur en chlorophylle pour la préconisation de doses d'azote. In: Guéréf M, King D, editors.Hétérogéneité parcellaire et gestion des cultures; vers une agricolture de précision. Editions QUAE. 2006. p. 179-198.
24. Houlès V, Mary B, Guérif M, Makovski D, Justes E. Evaluation of the ability of crop model STICS to recommend nitrogen fertilisation rates according to agro-environmental criteria. Agronomie 2004;24:339-349.
25. Jones JW, Hoogebroom G, Porter CH, Boote KJ, Batchelor WD, Hunt LA, Wilkens PW,Singh U, Gijsman AJ, Ritchie JT. The DSSAT cropping system model. European Journal of Agronomy 2003;18:235-265.
26. Jongschaap REE. Run-time calibration of simulation models by integrating remote sensing estimates of leaf area index and canopy nitrogen. European Journal of Agronomy 2006;24:316.
27. Justes E, Mary B, Meynard JM, Machet JM, Thelier-Huches L. Determination of a critical nitrogen dilution curve for winter wheat. Annals of Botanny 1994;74:297-407.
28. Kropff MJ, Bouma J, Jones JW. Systems approaches for the design of sustainable agro-ecosystemps. Agricultural Systems 2001;70:369-393.
29. Launay M, Guérif M. Ability of a model to predict crop production variability at the regional scale: evaluation for sugar beet. Agronomie 2003;23:135-146.
30. Lemaire G, Gastal F. N uptake and distribution in plant canopies. In: Lemaire G, editor. Diagnosis of the nitrogen status in crops. Berlin, Heidelberg: Springer-Verlag;1997.
31. Lemaire G, Gastal F, Salette J. Analysis of the effect of N nutrition on dry matter yield of a sward by reference to potential yield and optimum N content. Congrès International des Herbages, 1989;Vol. 1:179-180.

32. Makowski D, Wallach D, Tremblay M. Using Bayesian approach to parameter estimation; comparison of the GLUE and MCMC methods. Agronomie 2002;22:191-203.

33. Matthews R, Blackmore S. Precision agriculture 1997. Oxford, UK: Bios Scientific Publishers Ltd; 1997. Using crop simulation models to determine optimum management practices in precision agriculture; p. 413-420.

34. Moore GA, Tyndale-Biscoe JP. Estimation of the importance of spatially variable nitrogen application and soil moisture holding capacity to wheat production.Precision Agriculture 1999;1:27-38.

35. Moran S, Inoue Y, Barnes EM. Opportunities and limitations for image-based remote sensing in precision crop management. Remote Sensing of Enviroment1997;61:319-346.

36. Moulin S, Zurita Milla R, Guérif M, Baret F. *Characterizing the spatial and temporal variability of biophysical variables of a wheat crop using hyper-spectral measurements,* IGARSS 2003, Proceedings. Vol. 4. Toulouse, France: IEEE International; 2003. p. 2206-2208.

37. Nouvellon Y, Moran SM, Lo Seen D, et al. Combining a grassland ecosystem model with Landsat TM imagery for a ten-year simulation of carbon and water budget. Remote Sensing of Environment 2001;8:131-149.

38. Olioso A, Braud I, Chanzy A, et al. Monitoring energy and mass transfers during the Alpilles-ReSeDA experiment. Agronomie 2002;22:597-610.

39. Paloscia S, Macelloni G, Pampaloni P, Sigismondi S. The potential of C- and L- band SAR in estimating vegetation biomass: the ERS-1 and JERS-1 experiments. IEEE Transaction on Geoscience and Remote Sensing 1999;8:2107-2110.

40. Penuelas J, Gamon J, Freeden A, Merino J, Field C. Reflectance indices associated with physiological changes in nitrogen and water limited sunflower leaves. Remote Sensing of Environment 1994;48:135-146.

41. Pierce FJ, Nowak P. Aspects of precision agriculture. Advances in Agronomy1999;67:1-85.

42. Prévot L, Chauki H, Troufleau D, Weiss M, Baret F. Assimilating optical and radar data into the STICS model for wheat crops. Agronomie 2003;23:297-303.
43. Ruget F, Brisson N, Delécolle R, Faivre R. Sensitivity analysis of a crop simulation model, STICS, in order to choose the main parameters to be estimated. Agronomie2002;22:133-158.
44. Ruimy A, Saugier B, Dedieu G. Methodology for the estimation of terrestrial net primary production from remotely sensed data. Journal of Geophysical Research1994;99:5263-5283.
45. Tarentola A. Inverse problem theory methods for data fitting and model parameter estimation. Elsevier Science Publisher BV; 1987.
46. Ustin SL, Jacquemoud S, Zarco-Tejada PJ, Asner GP. Remote sensing of the environment: state of the science and new directions. In: Ustin SL, editor. Manual of remote sensing, *Vol. 4*. Remote sensing for natural resource management and environmental monitoring. John Wiley & Sons; 2004.
47. Verhoef W. Light scattering by leaf layers with application to canopy reflectance modeling: the SAIL model. Remote Sensing of Environment 1984;16:125-141.
48. Vouillot MO, Huet P, Boissard P. Early detection of N deficiency in wheat crop using physiological and radiometric methods. Agronomie 1998;18:117-130.
49. Weiss M, Baret F, Myneni R, Pragnère A, Knyazikhin Y. Investigation of a model inversion technique for the estimation of crop charcteristics from spectral and directional reflectance data. Agronomie 2000;20:3-22.
50. Wigneron J-P, Pardé M, Waldteufel P, Chanzy A, Kerr Y. Characterizing the dependence of vegetation model parameters on crop structure, incidence angle, and polarization at L-band. IEEE Transactions on Geoscience and Remote Sensing2004;42:416-425.

Chapter 2

Reducing Environmental Risk by Improving N Management in Intensive Chinese Agricultural Systems

Xiao-Tang Ju[a,1], Guang-Xi Xing[b], Xin-Ping Chen[a], Shao-Lin Zhang[b], Li-Juan Zhang[c], Xue-Jun Liu[a], Zhen-Ling Cui[a], Bin Yin[b], Peter Christie[a,d], Zhao-Liang Zhu[b] and Fu-Suo Zhang[a,1]

[a, 1]Key Laboratory of Plant and Soil Interactions, Ministry of Education, China, and College of Resources and Environmental Sciences, China Agricultural University, Beijing 100193, China.

[b]State Key Laboratory of Soil and Sustainable Agriculture, Institute of Soil Science, Chinese Academy of Sciences, Nanjing 210008, China.

[c]College of Agricultural Resources and Environmental Sciences, Hebei Agricultural University, Baoding 071001, China; and

[d]Agri-Environment Branch, Agri-Food and Biosciences Institute, Belfast BT9 5PX, United Kingdom.

ABSTRACT

Excessive N fertilization in intensive agricultural areas of China has resulted in serious environmental problems because of atmospheric, soil, and water enrichment with reactive N of agricultural origin. This study examines grain yields and N loss pathways using a synthetic approach in 2 of the most intensive double-cropping systems in China: waterlogged rice/upland wheat in the Taihu region of east China versus irrigated wheat/rainfed maize on the North China Plain. When compared with knowledge-based optimum N fertilization with 30–60% N savings, we found that current agricultural N practices with 550–600 kg of N per hectare fertilizer annually do not significantly increase crop yields but do lead to about 2 times larger N losses to the environment. The higher N loss rates and lower N retention rates indicate little utilization of residual N by the succeeding crop in rice/wheat systems in comparison with wheat/maize systems. Periodic waterlogging of upland systems caused large N losses by denitrification in the Taihu region. Calcareous soils and concentrated summer rainfall resulted in ammonia volatilization (19% for wheat and 24% for maize) and nitrate leaching being the main N loss pathways in wheat/maize systems. More than 2-fold increases in atmospheric deposition and irrigation water N reflect heavy air and water pollution and these have become important N sources to agricultural ecosystems. A better N balance can be achieved without sacrificing crop yields but significantly reducing environmental risk by adopting optimum N fertilization techniques, controlling the primary N loss pathways, and improving the performance of the agricultural Extension Service.

The last 40 years have seen an extraordinary change in the global N cycle (1). As recently as the 1960s, N availability in most parts of the world was controlled by natural processes, but the expanded production of synthetic N fertilizer and release of N from fossil fuel combustion now matches the natural rate of formation of reactive N on a worldwide basis (2). The rate of change has been spectacular with half of the synthetic N fertilizer ever used having been applied during the last 15 to 20 years (3). Furthermore, because of the difficulty in accurately predicting N fertilizer requirements, rates exceeding plant requirements are often applied, thus inducing unintended environmental consequences such as leaching of nitrate and emission of nitrous oxide and ammonia. This has become a major concern for scientists, environmental groups, and agricultural policymakers worldwide.

From 1977 to 2005, total annual grain production in China increased from 283 to 484 million tons (a 71% increase) and the average grain production per unit area increased from 2,348 to 4,642 (a 98% increase). However, synthetic N fertilizer application increased from 7.07 to 26.21 million tons (a 271% increase) over the same period. This resulted in a partial factor productivity from applied N (PEP_N) that decreased from 55 to 20 kg/kg (4, 5). Large amounts of external N inputs with decreasing N use efficiency have contributed to severe environmental degradation since the 1990s (6, 7). For example, the annual application rate of synthetic N for conventional agricultural practices in east and southeast China as well as the North China Plain now ranges from 550 to 600 kg of N per hectare for typical double-cropping systems (5, 8).

The Taihu region and the North China Plain are 2 of the most intensive agricultural regions in China and the most economically developed areas (5, 9). Large inputs of synthetic N fertilizer, rapid development of intensive livestock production systems, and rapidly increasing consumption of fossil fuels have severely disturbed regional biogeochemical N cycling and led to a series of environmental problems including eutrophication of surface waters, nitrate pollution of groundwater, acid rain and soil acidification, greenhouse gas emissions, and other forms of air pollution. There have also been effects on human health and normal functioning of ecosystems (9, 10), details of which are given in *SI Text*.

Here, we provide the actual grain yields and total N losses associated with knowledge-based optimum fertilization strategies compared with experience-based N management practices from numerous on-farm field experiments. The 'knowledge-based optimum N fertilization' (11) hereafter refers to using large numbers of field-based academic research results or soil tests on quantity and timing of synthetic N fertilizer practically, using regional mean optimal N (RMON) application rate in rice/wheat systems and in-season N management based on the soil mineral N test in wheat/maize systems (5, 8, 12, 13). We use an integrated approach to measure the fate of ^{15}N fertilizer and to document the major loss pathways *in situ* rather than by using individual measurements of loss processes reported in other studies to understand N behavior in these 2 rotations and thus attempt to control the environmental risk. We explain the reasons for the large external N inputs currently practiced in these intensive management systems and suggest strategies for balancing N management by applying knowledge-based optimum N fertilization techniques.

This study examines double-crop rotations, which are 2 of the most intensive agricultural systems worldwide and are widely practiced in Asian countries. The frequent alternation between flooding and draining the rice/wheat rotations led to significant changes in soil N transformations as compared with upland crop rotation. In addition, the extremely high external synthetic N inputs used in these two rotations during the last decade provided a unique opportunity to study the effects of N on environmental degradation. A detailed knowledge of N cycling and environmental impacts of these two intensive agricultural regions may also provide options for more rational N management practices in other intensive management cropping systems around the world.

Results and Discussion

We developed knowledge-based optimum N management techniques in rice/wheat systems in south China and wheat/maize system in north China during the last decade. The regional mean optimal N (RMON) application rate was calculated from the average of economically optimum N rates (i.e., the point of the N rate where the marginal grain production value is equal to the marginal N fertilization cost) based on large numbers of on-farm field experiments in the rice-based rotations and in-season N management based on the soil mineral N (Nmin: NH_4^+-N + NO_3^--N) test previously developed for upland crop rotations. In principle, RMON recommends a regional rational N rate in combination with fine tuning according to specific field conditions, taking into consideration the relative uniformity of climate, soil N fertility, and absence of a satisfactory soil N test index for rice-based rotations (5, 12). RMON gives the mean of economically optimum N rates and can be used as a reference for extension technicians. This recommendation method can be easily adapted to rural areas of China where soil testing facilities are usually not available and the cropping index is high. The short time interval between harvest of the first crop and sowing of the second makes soil testing impractical.

The improved Nmin method is based on synchronization of crop N demand and soil Nmin supply, and is considered to be a soil N supply index because of the high amounts of Nmin (mostly nitrate) frequently found in the root zone and its high availability for subsequent crop growth in the next phase of upland rotations. We found that high potential rates of mineralization and nitrification contribute to this high accumulation of nitrate in soils of the North China Plain (14). This method has been well established in wheat/maize rotations in recent years (8, 13, 15, 16) and can be easily adapted to individual fields because the nitrate test has become easy to perform in the laboratory and may even be feasible in the field (17). In the present study, we also evaluated the agronomic performance and potential environmental risk of these knowledge-based strategies in comparison with the farmer's experience-based practices.

We conducted numerous on-farm field experiments (Field Study 1) with different N levels in Taihu region and the North China Plain from 2003 to 2006. With the sole exception of wheat in Taihu region, the optimum N rate as determined using the regional mean optimal N application rate in Taihu region and the improved Nmin method on the North China Plain appeared to give slight increases in crop yield compared with farmers' current practices (Table 1). However, no significant differences ($P > 0.05$) were found, indicating that similar high crop yields can be achieved with lower N fertilizer rates because of increased soil indigenous N supply (18, 19). In contrast, total N losses (including NH_3 volatilization, denitrification and leaching from the top 1 m of the soil profile) increased significantly with increasing N inputs, indicating high environmental costs were caused by exceeding optimum N fertilizer rates.

Table 1: Average grain yields and total N losses of the optimum N fertilization (ON) compared with farmers' N practices (FN) (Field Study 1 and 2).

Crop and site of field experiment	N fertilization	N rate		Grain yield		Total fertilizer N loss*	
		Rate, kg of N per hectare	Ratio of FN to ON	Yield, $kg \cdot ha^{-1}$	Ratio of FN to ON	Total loss, kg of N per hectare	Ratio of FN to ON
Rice in Taihu (*n* = 26)	ON†	200		8,270		102	
	FN	300	1.5	8,012	0.97	174	1.7
Wheat in Taihu (*n* = 9)	ON†	153		3,700		76	
	FN	250	1.6	4,084	1.10	155	2.0
Wheat in NCP‡(*n* = 121)	ON§	128		6,024		25	
	FN	325	2.5	5,764	0.96	71	2.8
Maize in NCP‡(*n* = 148)	ON§	158		8,900		52	
	FN	263	1.7	8,500	0.95	108	2.1

*Total fertilizer N losses calculated with the models of Fig. 1B simulated from ^{15}N field experiments.

†Regional mean optimal N application rate calculated from the mean of economically optimum N rates of field experiments in Taihu regain (5, 12).

‡Data including Field Study 1 and also summarized from ref. (15, 16); NCP, North China Plain.

§In-season nitrogen management based on soil Nmin test on the NCP (8, 13, 15, 16).

We used ^{15}N in some of the field experiments (Field Study 2) to quantify differences in fertilizer N fate at different application rates among the four crops. We found that N recovery rates decreased with increasing N application rate for all of them (Fig. 1*A*). N rates showed polynomial, linear, polynomial, and logarithmic relationships with N recovery by rice, wheat (south), wheat (north) and maize, respectively. Each fitting produced a highly significant model ($P < 0.01$). At lower N input levels (<125 kg of N per hectare), the order was: wheat (north) > maize > wheat (south) > rice. At higher N input levels (>125 kg of N per hectare) the order was: wheat (north) > rice > maize > wheat (south), indicating that rice recovered more N at high N application rates. N loss rates increased with increasing N application rate in all four crops (Fig. 1*B*). A logarithmic model satisfactorily simulated the correlations between N rate and N loss rates for the four crops ($P < 0.01$). Rice and wheat (south) in the Taihu region showed much higher N loss rates than did wheat (north) and maize on the North China Plain. The trends of the N retention rates with increasing N rate differed among the four crops.

The rice and wheat systems in the Taihu region showed lower N retention rates than wheat and maize systems on the North China Plain. Moreover, N retention rates in wheat (south) and wheat (north) decreased first and then increased with further increase in N rate, showing typical polynomial relationships ($P < 0.01$). However, N retention rates in rice and maize decreased by power function relationships when N rate increased ($P < 0.01$). These findings indicate that the upland wheat and maize rotation system has significantly higher N recovery rates, higher N retention rates, and lower N loss rates than the waterlogged rice and upland wheat rotation in the Taihu region under current soil N fertility and farmers' management practices. Moreover, on the North China Plain wheat has significantly higher N recovery rates, higher N retention rates, and lower N loss rates than maize, and in the Taihu region, rice has significantly higher N recovery rates and lower N loss rates than wheat. Rice/wheat rotations show high 'leakage' of external N inputs with resulting pollution threats to water and air. However, wheat/maize rotations show strong retention of external N inputs that can remain available for use by subsequent crops. Retention of N is favorable for long-term improvement of soil N fertility and further emphasizes the need to reduce N application rates. It should be noted that N rates applied to wheat in the Taihu region are similarly high to those on the North China Plain, but wheat grain yields are much lower than in the north and show much lower N recovery rates (Table 1). One explanation is that farmers in the Taihu region try to compensate for the unfavorable growing conditions for wheat (e.g., extremely wet conditions in the root zone due to high rainfall and groundwater) by increasing N fertilizer application rates (20).

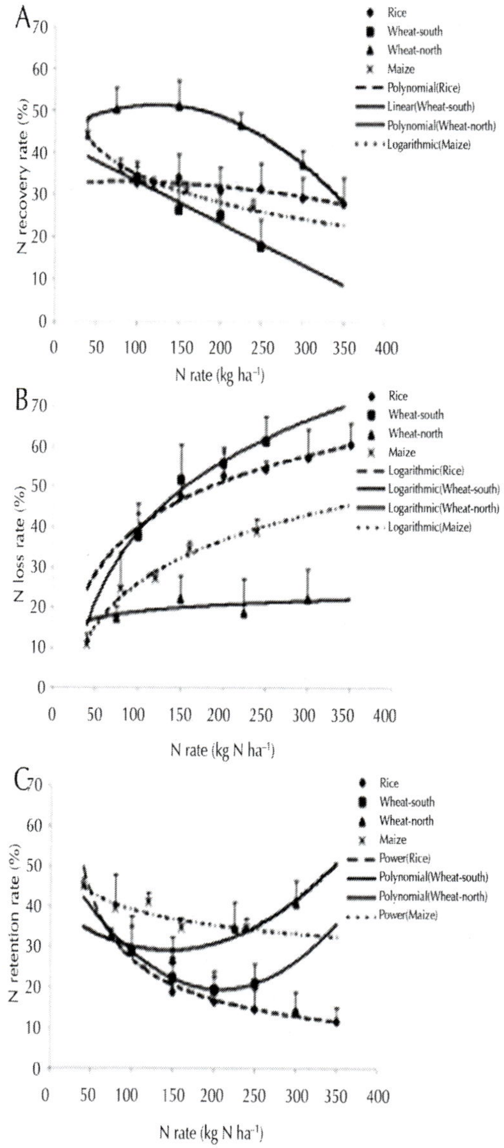

Figure 1: Relationships of N recovery rate (*A*), loss rate (*B*), and retention rate (*C*) with N application rate in 4 crops (Field Study 2). Vertical bars denote standard deviation of the mean (average of 6 ^{15}N field experiments). Each model fitting produced a highly significant model ($P < 0.01$).

To further understand the different pathways of N loss from current N practices and to control them in the future, we conducted a lysimeter study (Lysimeter Study) in the Taihu region and a 4-year field experiment (Field Study 3) on the North China Plain, measuring in situ NH_3 volatilization, denitrification, N leaching, and N_2O emissions. We found that these N loss pathways differed sharply among the four crops because of differences in climate, soils and management practices (Table 2). Ammonia volatilization seems to be the main loss pathway for fertilizer N, with losses accounting for 19.4 and 24.7% of applied N in wheat (north) and maize seasons, respectively. Volatilization was much lower for rice and wheat (south), with losses accounting for 11.6 and 2.1% of applied N, respectively. Denitrification losses were quite small in the wheat and maize on the North China Plain (0.1 and 3.3% of applied N). In contrast, denitrification was the primary pathway for fertilizer N loss in rice and wheat within the Taihu region, accounting for 36.4 and 43.5% of applied N, respectively. Leaching accounted for only 0.3% of applied N from rice and 3.4% from wheat (south) season, but accounted for 2.7 and 12.1% of applied N for wheat and maize growing seasons on the North China Plain. We may be overestimating the denitrification loss in rice/wheat systems due to error accumulation with the different methods (5), but the order of magnitude agrees with previous studies (5).

Table 2: Different N loss pathways expressed as a percentage (mean ± SD) of N application rate in farmers› N practices (Field Study 3, Lysimeter Study)

Component		**Taihu region**		**North China Plain**	
		Rice	**Wheat-south**	**Wheat-north**	**Maize**
N rate (kg of N per hectare)		300	250	325	263
Recovery rate (%)*		29.6 ± 4.9	18.4 ± 6.3	31.0 ± 3.6	25.5 ± 5.2
Retention rate (%)*		21.7 ± 5.1	28.5 ± 4.6	45.7 ± 5.4	33.9 ± 2.3
Loss pathway	NH_3 volatilization (%)	11.6 ± 4.7	2.1 ± 1.4	19.4 ± 5.2	24.7 ± 5.6

	Leaching out of 1 m soil depth (%)	0.3 ± 0.5	3.4 ± 2.1	2.7 ± 2.6	12.1 ± 8.5
	Denitrification (%)	36.4†	43.5†	0.1 ± 0.04	3.3 ± 1.6

*Measured from corresponding ^{15}N field experiments.

†Calculated by difference method.

Alternating upland wheat and waterlogged rice led to high denitrification of applied N and accumulated nitrate (after wheat harvest) in the rice season (20–22). The traditional practice of surface application of urea-N stimulated NH_3 volatilization from rice (23, 24). After plowing of paddy soil, a hard layer that formed below the plow layer due to clay deposition appeared to reduce leakage from the rice crop, but resulted in higher leaching losses during the wheat season when drainage prevails (25). The high denitrification loss for wheat (south) is attributed to wet soil conditions, higher temperatures and relatively high soil carbon (C) content in the Taihu region (26, 27). During the wheat season in the Taihu region, N fertilizer (urea) is readily nitrified and accumulates in the soil, consequently becoming available for loss through denitrification during the rice season (22). Thus the residual NO_3^--N in the soil profile is not fully available to the subsequent crop as is the case in upland crop rotations. We speculate that the very large N losses (corresponding with lower N retention rates) and large C inputs in straw under the anaerobic conditions in rice/wheat rotations may be another explanation for the structural changes in soil organic matter and increased C-rich humic acids found by Schmidt-Rohr *et al.* (28). On the North China Plain, with its calcareous soils, typical pH 7.5 to 8.5 and predominant use of urea or ammonium bicarbonate as fertilizer (accounting for 80%), NH_3 volatilization is generally considered to be the major pathway of N loss, especially in the hot maize growing season (29–31). Due to lower soil water content and soil organic matter (1.0–1.5%), most experimental results on the North China Plain have shown the magnitude of N loss by denitrification is no more than 10 kg of N per hectare and that denitrification accounts for less than 5% of the fertilizer N, most of which occurs during the hot and humid summer season (30, 32). We found that nitrate accumulation in the soil profile and gradual leaching below the root zone were the important N loss pathways in wheat and maize rotations under conventional agricultural N management practices due to concentrated rainfall in summer and flood irrigation (8, 33), contradicted conventional thinking that leaching losses are not an important pathway for N loss in semi-humid upland agricultural systems of the North China Plain.

Field Study 1 and the Lysimeter Study also showed that NH_3 volatilization increased rapidly and represented the major pathway for loss in response to increasing N rate, but N_2O emission and leaching did not increase dramatically with increasing N rate during the rice growing season (Fig. S1, rice). Leaching did increase rapidly and accounted for most of the reactive N loss with increasing N fertilizer rate, while N_2O emissions and NH_3 volatilization increased very little in response to N rate during the wheat season in the Taihu region (Fig. S1, wheat-south). Ammonia volatilization and NO_3-N leaching increased significantly with increasing N rate and N_2O emissions also showed an upward trend in the wheat growing season on the North China Plain (Fig. S1, wheat-north). During the maize growing season, NH_3 volatilization, NO_3-N leaching, and N_2O emissions all increased significantly, with N leaching showing a particularly strong increase (Fig. S1, maize). These results confirm that emissions of reactive N to the environment will increase after N fertilization exceeds the optimum rate. However, we did not find large N_2O emissions among the four crops. The total N_2O emissions in rice, wheat (south), wheat (north), and maize with different N application rates were 0.13 to 0.50, 0.27 to 0.63, 0.13 to 0.24, and 0.16 to 0.65 kg of N per hectare, respectively. Expressed as average emission factors, this accounted for 0.20, 0.30, 0.12, and 0.23%, respectively, close the IPCC 2006 guideline of 0.3% for paddy rice soil and much lower than the guideline of 1% for upland soil (27, 34, 35). We explain these results by the low ratio of N_2O to N_2O+N_2 from denitrification products (36, 37) under waterlogged rice conditions, low available carbon sources and low water-filled pore space (WFPS) for denitrification in most semi-humid upland soils on the North China Plain (14).

To quantify atmospheric and irrigation water contributions (hereafter 'environmental N inputs') to agricultural systems in the Taihu region and the North China Plain, we measured deposition and N inputs from irrigation water (Monitoring Study). Total environmental N inputs reached about 89 kg of N per hectare in the Taihu region and about 104 kg of N per hectare on the North China Plain. Comparisons with observations made from both regions in the 1980s shows more than 2-fold greater environmental N inputs with time. N deposition increased from 15 to 33 kg of N per hectare and irrigation water N inputs from 15 to 56 kg of N per hectare in Taihu region (Fig. S2) (38). During the same period, N deposition increased significantly from 21 to 89 kg of N per hectare and irrigation water N inputs from 8 to 15 kg of N per hectare (Fig. S2) (38, 39) on the North China Plain. Inorganic N and soluble organic N are the primary environmental N sources that can be used directly by crops or after transformation in the soil (40). The absence of crop yield response to N application in most field experiments since the 1990s (5, 8, 15, 16, 20) is not only attributable to increased indigenous soil N supply, but also to high environmental N inputs (41, 42). Moreover, high N deposition may also contribute to surface water eutrophication, acid rain, and soil acidification (43) in these regions. High total N concentrations (7 to 8 $mg{\cdot}N{\cdot}L^{-1}$) in the irrigation surface water in the Taihu region and high NO_3^--N concentrations in the irrigation shallow groundwater on the North China Plain also reflect the heavy N pollution of water resources in both areas.

We calculated annual N balance for both rotations using two scenarios: conventional N practice versus optimum N fertilization (Tables S1 and S2). We computed an annual N surplus of 87 kg of N per hectare for current practices with large losses by denitrification for the rice/wheat system. Synthetic N fertilizer was the primary input followed by N in irrigation water, biological N_2 fixation, and N deposition. A better N balance can be achieved by adopting optimum N fertilization strategies designed to maintain relatively high yields but reduce environmental risk. However, dinitrification could be further reduced by improving carbon management and controlling the water regime (25–27). We also computed a 212 kg of N per hectare surplus for current practice with large losses by NH_3 volatilization within the wheat/maize system. A large proportion of surplus N accumulated as nitrate in the soil profile after harvest (33), and partly existed as N in organic form due to manure application. However, the unusually large quantity of annual surplus N might also be caused by underestimating the leaching loss due to drought conditions in our observation years with 24–46% reduction in rainfall (8). Previous studies showed that strong N leaching losses only occurred in some years with heavy summer rainfall (33), leading to high nitrate accumulation in the deep subsoil

(8, 33) and the groundwater (43). In the future, a slightly negative N balance could be achieved using an optimum N rate of about 286 kg of N per hectare to maintain relatively high yields (15, 16). The slightly negative balance would be conducive to making sure the plants fully use accumulated nitrate and further reduce nitrate leaching (8, 33). A high N retention rate is conducive to maintenance of soil N fertility. Changing the N application techniques and using deep placement of urea or ammonium bicarbonate could substantially reduce NH_3 volatilization losses from calcareous soils (29).

A question that needs to be addressed is why farmers in China continue to use such excessively high rates of N fertilizer. Before the 1990s, scientists, government and extension staff encouraged farmers to increase synthetic fertilizer inputs to increase yields and feed an increasing population. As a result, synthetic fertilizer inputs increased continually throughout the country, especially in the Taihu region and North China Plain. By 2000, the rates were far greater than crop demand and serious environmental degradation had begun. Although this trend has been recognized by the scientific community since the late 1990s, on-farm practices are difficult to reverse after 10 to 15 years of effort (43). Persuading farmers to limit fertilizer inputs is difficult because many of them still hold to now traditional opinions that higher crop yield will be obtained with more fertilizer. Application rates of N therefore often include an extra 'insurance' component to prevent yield loss rather than matching inputs to crop demand. The high off-farm incomes and relatively low retail prices for N fertilizers (with government subsidies for production and transportation) compared with U.S. and European prices are also important factors (43, 44). A third major reason is the poor infrastructure of the Extension Services and poor transfer of knowledge to farmers (43). A critical objective of the optimum N techniques developed using our studies are to substantially lower fertilizer N application rates by accounting for indigenous soil N and environmental N inputs for maintenance of the yields needed to feed an increasing population. This strategy also limits losses of total N to the water and atmosphere with the aim of establishing sustainable agricultural systems (8, 11, 12).

Crop production in intensive agricultural practice currently relies too heavily on synthetic N fertilizer inputs and cereal monocropping, especially in Asian countries under pressure to feed large and growing populations. It is now time to change this situation by balancing yield and environmental consequences. Integrated management packages need to be developed for major cropping systems that include efficient recycling of manures and crop residues, the use of legume crops in rotations to increase internal N cycling and further reduction in the reliance on synthetic N fertilizers (45).

CONCLUSIONS

It took nearly 50 years to achieve food sufficiency in China. An unanticipated cost has been that massive fertilizer inputs have led to significant environmental degradation (6). Over-fertilization is a serious problem in intensive agricultural production areas in China, resulting in enrichment of reactive N in the air, soil and water with consequent impairment of ecosystem services. Our studies show that more efficient use of N fertilizer can allow current N application rates to be reduced by 30 to 60%. This would still maintain crop yields and N balance in rotations, while substantially reducing N losses to the environment. The over-application of N also represents an unnecessary economic expenditure for farmers. The new recommendations should fully take into account the N supplying capacity of the soil and N deposited from air and irrigation water. The characteristics of N behavior among the 4 crops were sharply different depending on climatic, soil and management practices. This must be taken into consideration to further reduce N losses. Only by reducing fertilizer N inputs can degraded environments be gradually restored, enhanced and protected. Although several environmental standards have been set in the past, there are still no legislative controls in China equivalent to those in the European Union (33). China would benefit from adopting and enforcing relevant agricultural regulations (7). All these goals could be achieved by removing government subsidies, introducing an N fertilizer tax, improving local Extension Services, educating farmers for environmental awareness, and employing practices that avoid serious environmental degradation (43).

Materials and Methods

Study Areas. Two different representative intensive agricultural regions were selected as study areas: (1) the Taihu region in east China (30–32

°N, 119–122 °E) in the alluvial delta of the Yangtze River and (2) the North China Plain in northeast China (32–41 °N, 113–120 °E) in the alluvial plain of the Yellow River (Fig. S3). Details of the climate, soils, and crops are given in *SI Text*.

Field Study 1. In the Taihu region, 26 rice (8 or 9 locations each year) and 9 wheat (3 locations each year) on-farm field experiments with 7 (rice) or 5 (wheat) N rates were conducted from 2003 to 2006. The 7 rice N treatments were 0, 100, 150, 200, 250, 300, and 350 kg of N per hectare and the 5 wheat N treatments were 0, 100, 150, 200, and 250 kg of N per hectare. The plots (42 m^2 in area, 7 × 6 m) were arranged in a randomized complete block experimental design with 4 replicates.

On the North China Plain 6 winter wheat and 6 summer maize on-farm field experiments with 5 (winter wheat) or 6 (summer maize) N rates were conducted from October 2003 to October 2005 at Dongbeiwang near Beijing, Huimin County in Shandong province, and Baoding County in Hebei province. The 5 winter wheat N treatments were 0, 75, 150, 225, and 300 kg of N per hectare and the 6 summer maize N treatments were 0, 40, 80, 120, 160, and 240 kg of N per hectare. The plot size was 63 m^2 (9 × 7 m). The plots were arranged in a randomized complete block experimental design with 3 replicates. The fertilizer types and application times are given in the *SI Text*. Except for fertilizer application and grain harvest, each experimental field was managed using the individual farmer's current management practices in both rotations. Aboveground biomass was measured by hand in all plots. The central area of 16 m^2 (4 × 4 m) in each plot was harvested to determine grain dry matter yield.

Field Study 2. In the Taihu region, 6 rice (2 locations per year) and 6 wheat (2 locations per year) field experiments were selected to conduct ^{15}N studies in microplots to determine the fate of N fertilizer at different N levels. On the North China Plain, the ^{15}N study was carried out in microplots in the on-farm field experiments (*SI Text*). In both rotations plant and soil sampling and analysis for total N and ^{15}N abundance are described as in (44). The N recovery rate is expressed as the percentage of applied ^{15}N fertilizer taken up by the aboveground plant parts and the N retention rate as the percentage of applied ^{15}N fertilizer recovered in the top 100 cm of the soil profile. The loss rate was calculated by subtracting the recovery rate and retention rate from

100. The mean and standard variation was calculated across all of the experimental years and sites in same crop species.

Lysimeter Study. Twenty-four lysimeters containing undisturbed soil profiles were used at the research station at Changshu, Jiangsu province (31° 31.93′ N, 120° 41.88′ E) in the Taihu region. They contained a Typic Epiaquept (46) formed on alluvial loess with a silty clay loam (46) texture. A steel cylinder 1-m deep with 1.14-m inner diameter was pushed into the soil, cut at the base, and then removed. After collection, the lysimeters were prepared for free drainage by replacing approximately 0.08 m of soil at the base of each column with a nylon mesh, a gravel layer, and a porous plastic sheet (47). The lysimeters were then placed in plots permanently installed belowground at the station. Each lysimeter was used to grow a waterlogged summer rice/upland winter wheat rotation from October 2003 to October 2005 with 2 rice and 2 wheat growing seasons. Details of treatments are given in *SI Text*. Water leaching through the soil columns was collected in plastic containers placed in an underground measuring station. The leachate volume was measured at 3-day intervals and subsamples were taken from the accumulated leachate and stored at −18 °C before chemical analysis. During the winter wheat growing seasons, leachate could be collected only during rainfall periods. Leachate NO_3^-, NH_4^+, and total N concentrations were determined using a continuous flow analyzer (TRAACS 2000, Bran and Luebbe Inc.). The amount of N leached at each collection time was calculated by multiplying the volume and the N concentration of leachate. The total amount of N leached over the whole growing period was obtained by summing the individual amounts of N leached at the different collection times. The N leached from fertilizer was calculated as the difference in leached N between fertilized treatments and unfertilized controls. The mean and standard variation of each N treatment was calculated across 3 replicates over 2 years.

NH_3 volatilization and N_2O measurements were conducted in both rice and both wheat growing seasons from October 2003 to October 2005. Ammonia volatilization rate was measured with a continuous airflow enclosure method (48). The total amount of NH_3 volatilized over the whole growing period was obtained by summing NH_3 volatilization measured each day. N_2O fluxes were usually measured once a week throughout each crop growing period with static closed chambers, and measurements were made more frequently after basal fertilization and topdressing. Detail measurements and calculations are described in (49). Similar NH_3-volatilization and N_2O measurements at different N levels were conducted in the on-farm field experiment at Baoding County in Hebei province (Field Study 1). The amount of fertilizer derived NH_3 volatilization or N_2O emission was calculated as the difference between volatized NH_3 or N_2O in fertilized and unfertilized treatments. The mean and standard variation of each N treatment was calculated across 3 replicates over 2 years.

Field Study 3. A field experiment was conducted from September 2000 to September 2004 at Dongbeiwang research station near Beijing (40.08 °N, 116.28 °E) on a calcareous alluvial soil [Fluvaquents (46)] considered typical of the North China Plain. Eight successive crops (4 winter wheat and 4 summer maize) were grown on the same plots over the 4-year period. Detailed descriptions of the field experiment were reported previously (8).

^{15}N studies in microplots were also conducted in the conventional N and optimized N fertilization treatments from September 2002 to September 2004. NH_3 volatilization was measured from September 2002 to September 2004 with an automatic wind tunnel system that covered an experimental soil area of 0.7 × 1.5 m (1.05 m^2) (50). Field denitrification was monitored by a soil core incubation system using the C_2H_2-inhibition method (51) during the 2-year study from September 2002 to September 2004 (32). Nitrate leaching was quantified using TerrAquat passive samplers filled with ion-exchange resin. These were used to quantify the total annual amount of leached nitrate below the rooting zone with ten replicates per plot (52). The mean and standard variation of NH_3, N_2O, and nitrate leaching were calculated across 4 replicates over 2 years.

Monitoring Study. Measurements of wet and dry deposition were conducted from October 2003 to October 2005 in the 3 representative locations of the Taihu region, that is, Changshu, Nanjing, and Hangzhou. Three automatic wet-only samplers (APS series, Wuhan Tianhong Inc.), which collected precipitation samples only during precipitation events (controlled by sensors) were separately installed at the above 3 sites for the collection of wet-only deposition (53). Particulate dry N samples were collected using a 0.5-m^2 polyethylene sheet-based sampler (41). Before sunrise (about 6:00 AM in winter and around 5:00 AM in summer), 250 mL of 2 M KCl solution was used to collect deposited dew and the particles in the collector. In the present experiment, the N in dry deposition is defined as only the part of the inorganic N adsorbed to particles and exchangeable with 2 M KCl.

Precipitation samples were collected from 11 monitoring sites located on the North China Plain during the period 2003–2006. Bulk deposition (wet deposition plus sedimenting dry deposition such as dust during dry periods) was collected (41, 53). The mean of the bulk deposition was calculated across all of the sites over the measuring years. Total deposition was determined by the ITNI system at Dongbeiwang, Beijing and Wuqiao and Quzhou in Hebei province. The ITNI system is based on ^{15}N isotope dilution and allows the inclusion of direct N uptake by the plant as well as the absorption of gaseous N compounds by the soil-plant system (40, 54, 55).

The N inputs from irrigation were calculated from the sum of the amounts of water in the periods of irrigation and the corresponding concentrations of ammonium and nitrate in the irrigation water. The measurements were conducted from October 2003 to October 2005 at 12 locations in the Taihu region and 10 locations on the North China Plain. The mean value was calculated across all of the years and sites.

Statistical Analysis. Most of the data are presented as mean ± SD (standard deviation) in the Tables and Figures. All data were analyzed using the SAS software package (SAS Institute). Means were compared using LSD (Least Significant Difference) by one-way ANOVA analysis. Regression analysis of N fertilizer rate vs. recovery rate, retention rate, and loss rate was performed using the PROC REG procedure.

Acknowledgments

We thank Professor D. Tilman and three anonymous reviewers for their valuable comments and suggestions that have greatly improved the manuscript. This work was supported by the National Natural Science Foundation of China (Grant Nos 30390080, 30821003, 40771098, & 40571072), BMBF Germany No. 0339712A and the Program for Changjiang Scholars and Innovation Research Teams in Universities (IRT 0511).

REFERENCES

1. Galloway JN, et al. (2008) Transformation of the nitrogen cycle: Recent trends, questions, and potential solutions. Science **320**:889–892.
2. Galloway JN, et al. (2004) Nitrogen cycles: Past, present, and future. Biogeochemistry**70**:153–226.
3. Erisman JW, Bleeker A, Galloway JN, Sutton MS (2007) Reduced nitrogen in ecology and the environment. Environ Pollut **150**:140–149.
4. Editorial Committee of China Agricultural Yearbook *China Agricultural Yearbook 1980–2006 (in Chinese)* (China Agricultural Press, Beijing).
5. Zhu ZL, Chen DL (2002) Nitrogen fertilizer use in China - contributions to food production, impacts on the environment and best management Strategies. Nutr Cycl Agroecosyst **63**:117–127.
6. Liu JG, Diamond J (2005) China's environment in a globalizing world. Nature **435**:1179–1186.
7. Liu JG, Diamond J (2008) Revolutionizing China's environmental protection. Science **319**:37–38.
8. Zhao RF, et al. (2006) Fertilization and nitrogen balance in a wheat-maize rotation system in North China. Agron J **98**:938–945.
9. Zhu ZL, Xiong ZQ, Xing GX (2005) Impacts of population growth and economic development on the nitrogen cycle in Asia. Sci China C life Sci **48**:729–737.
10. Jin XC, Xu QJ, Huang CZ (2005) Current status and future tendency of lake eutrophication in China. Sci China C life Sci **48**:948–954.
11. Matson PA, Naylor R, Ortiz-Monasterio I (1998) Integration of environmental, agronomic, and economic aspects of fertilizer management. Science **280**:112–115. Zhu ZL
12. (2006) On the methodology of recommendation for the application rate of chemical fertilizer nitrogen to crops (in Chinese with English abstract) Plant Nutri Fert Sci **12**:1–4.

13. Chen XP, et al. (2006) Synchronizing N supply from soil and fertilizer and N demand of winter wheat by an improved Nmin method. Nutr Cycl Agroecosyst **74**:91–98.

14. Wan YJ, et al. (2008) Gross nitrogen transformations and related N_2O emissions in an intensively used calcareous soil. Soil Sci Soc. Am. J.

15. Cui ZL, et al. (2008) On-farm evaluation of an in-season nitrogen management strategy based on soil Nmin test. Field Crops Res **105**:48–55.

16. Cui ZL, et al. (2008) On-farm evaluation of the improved soil N_{min}-based nitrogen management for summer maize in North China Plain. Agron J **100**:517–525.

17. Schmidhalter U (2005) Development of a quick on-farm test to determine nitrate levels in soil. J Plant Nutr Soil Sci **168**:1–7.

18. Cui ZL, et al. (2008) On-farm estimation of indigenous nitrogen supply for site-specific nitrogen management in the North China Plain. Nutr Cycl Agroecosyst **81**:37–47.

19. Peng SB, et al. (2002) Research strategy in improving fertilizer nitrogen use efficiency of irrigated rice in China (in Chinese with English abstract) Agri Sci China (Chinese version) **35**:1095–1103.

20. Roelcke M, et al. (2004) Recent trends and recommendations for nitrogen fertilizer in intensive agriculture in Eastern China. Pedosphere **14**:449–460.

21. Richter J, Roelcke M (2000) The N-cycle as determined by intensive agriculture— examples from Central Europe and China. Nutr Cycl Agroecosyst **57**:33–46.

22. Fan MS, et al. (2007) Nitrogen input, ^{15}N balance, and mineral N dynamics in a rice-wheat rotation in southwest China. Nutr Cycl Agroecosyst **79**:255–265.

23. Lin DX, Fan XH, Hu F, Zhao HT, Luo JF (2007) Ammonia volatilization and nitrogen utilization efficiency in response to urea application in rice fields of the Taihu lake region, China. Pedosphere**17**:639–645.

24. Cai GX, et al. (1992) Gaseous loss of nitrogen from fertilizers applied to a paddy soil in southeastern China. Pedosphere **2**:209–217.

25. Xing GX, Zhu ZL (2000) An assessment of N loss from agricultural fields to the environment in China. Nutr Cycl Agroecosyst **57**:67–73.
26. Xing GX, et al. (2002) Denitrification in underground saturated soil in a rice paddy region. Soil Biol Biochem **34**:1593–1598.
27. Xiong ZQ, et al. (2002) Measurement of nitrous oxide emissions from two rice-based cropping systems in China. Nutr Cycl Agroecosyst **64**:125–133.
28. Schmidt-Rohr K, Mao JD, Olk DC (2004) Nitrogen-bonded aromatics in soil organic matter and their implications for a yield decline in intensive rice cropping. PNAS **101**:6351–6354.
29. Cai GX, Fan XH, Yang Z, Zhu ZL (1998) Gaseous loss of nitrogen from fertilizers applied to wheat growing on a calcareous soil in North China Plain. Pedosphere **8**:45–52.
30. Zhang YM, et al. (2004) Ammonia volatilization and dinitrification losses from an irrigated maize-wheat rotation field in the North China Plain. Pedosphere **14**:533–540.
31. Zhang SL, et al. (1992) Losses of urea-nitrogen applied to maize on a calcareous fluvo-aquic soil in North China Plain. Pedosphere **2**:171–178.
32. Zou GY, Zhang FS, Ju XT, Chen XP, Liu XJ (2006) Study on soil denitrification in wheat-maize rotation system. Agri Sci China (English version) **5**:45–49.
33. Ju XT, Liu XJ, Zhang FS, Roelcke M (2004) Nitrogen fertilization, soil nitrate accumulation, and policy recommendations in several agricultural regions of China. Ambio **33**:300–305.
34. Akiyama H, Yagi K, Yan X (2005) Direct N_2O emissions from rice paddy fields: Summary of available data. Global Biogeochem Cycles , 19. GB1005.
35. IPCC (2006) *IPCC 2006 Guidelines for National Greenhouse Gas Inventories Volume 4: Agriculture, Forestry, and Other Landuse* (OECD, Paris), pp 475–505.
36. Firestone M, Firestone R, Tiedje J (1980) Nitrous oxide from soil denitrification: Factors controlling its biological production. Science **208**:749–751.

37. Weier KL, Doran JW, Power JF, Walters DT (1993) Denitrification and the dinitrogen/nitrous oxide ration as affected by soil water, available carbon and nitrate. Soil Sci SocAmJ **57**:66–72.
38. Zhu ZL (1997) in *Nitrogen in Soils of China*, Nitrogen balance and cycling in agroecosystems of China, eds Zhu ZL, Wen QX, Freney JR (Kluwer Academic Publishers, Dordrecht/Boston/London), pp323–338.
39. Shen SM (1998) in *Soil Fertility in China*, eds Shen SM, Wan HF, Xie JC (China Agricultural Press,Beijing), pp 64–80.
40. He CE, Liu XJ, Fangmeier A, Zhang FS (2007) Quantifying the total airborne nitrogen input into agroecosystems in the North China Plain. AgricEcosystem Env **121**:395–400.
41. Zhang Y, Liu XJ, Fangmeier A, Goulding KTW, Zhang FS (2008) Nitrogen inputs and isotopes in precipitation in the North China Plain. Atmos Env **42**:1436–1448.
42. Zhang Y, et al. (2008) Evidence for organic N deposition and its anthropogenic sources in China.Atmos Env **42**:1035–1041.
43. Zhu ZL, Norse D, Sun B (2006) *Policy for Reducing Non-point Source Pollution from Crop Production in China* (China Env. Sci. Press, Beijing), pp 135–299.
44. Ju XT, Liu XJ, Pan JR, Zhang FS (2007) Fate of ^{15}N-labeled urea under a winter wheat/summer maize rotation on the North China Plain. Pedosphere **17**:52–61.
45. Ju XT, Zhang FS, Bao XM, RömheldV, Roelcke M (2005) Utilization and management of organic wastes in Chinese agriculture: Past, present, and perspectives. Sci China C life Sci **48**:965–979.
46. United States Department of Agriculture (1994) Keys to Soil Taxonomy 6th Ed. (Soil Conservation Service).
47. Persson L, Bergström L (1991) A drilling method for collection of undisturbed soil monoliths. Soil Sci Soc AmJ **55**:285–287.
48. Tian GM, Cao JL, Cai ZC, Ren LT (1998) Ammonia volatilization from winter wheat field top-dressed with urea. Pedosphere **8**:331–336.
49. Xiong ZQ, et al. (2002) Field study on nitrous oxide emissions from upland cropping systems in China. Soil SciPlant Nutri **48**:539–546.

50. Braschkat J, Mannheim T, Horlacher D, Marschner H (1993) Measurement of ammonia emissions after liquid manure application: Construction of a wind tunnel system for measurement under field conditions. Z Pflanzenernahr Bodenk **156**:393–396.

51. Ryden JC, Skinner JH, Nixon DJ (1987) Soil core incubation system for the field measurement of denitrification using acetylene-inhibition. Soil Biol Biochem **19**:753–757.

52. Mack UD, Feger KH, Gong YS, Stahr K (2005) Soil water balance and nitrate leaching in winter wheat-summer maize double-cropping systems with different irrigation and N fertilization in the North China Plain. J Plant Nutr Soil Sci **168**:454–460.

53. Liu XJ, et al. (2006) Nitrogen deposition in agroecosystems in the Beijing area. AgricEcosystem Env **113**:370–377.

54. Russow R, Boehme F, Neue HU (2001) A new approach to determine the total airborne N input into the soil/plant system using ^{15}N isotope dilution (ITNI): Results for agricultural areas in Central Germany. Science World **1**:255–260.

55. Russow R, Boehme F (2005) Determination of the total nitrogen deposition by the ^{15}N isotope dilution method and problems in extrapolating results to field scale. Geoderma **127**:62–70.

Chapter 3

Modelling Cereal Root Systems for Water and Nitrogen Capture: Towards an Economic Optimum

John King[1], Alan Gay[2], Roger Sylvester Bradley[1], Ian Bingham[3], John Foulkes[4], Peter Gregory[5], and David Robinson[6]

[1]ADAS Ltd, Boxworth Research Centre, Boxworth, Cambridge CB3 8NN, UK

[2]IGER, Plas Gogerddan, Aberystwyth, Ceredigion SY23 3EB, UK

[3]Scottish Agricultural College, Craibstone Estate, Aberdeen AB21 9YA, UK

[4]Department of Agriculture and Horticulture, University of Nottingham, Sutton Bonington Campus, Loughborough LE12 5RD, UK

[5]Department of Soil Science, University of Reading, PO Box 233, Whiteknights, Reading RG6 6DW, UK

[6]Department of Plant and Soil Science, University of Aberdeen, Aberdeen AB24 3UU, UK

ABSTRACT

A quantitative model of wheat root systems is developed that links the size and distribution of the root system to the capture of water and nitrogen (which are assumed to be evenly distributed with depth) during grain filling, and allows estimates of the economic consequences of this capture to be assessed. A particular feature of the model is its use of summarizing concepts, and reliance on only the minimum number of parameters (each with a clear biological meaning). The model is then used to provide an economic sensitivity analysis of possible target characteristics for manipulating root systems. These characteristics were: root distribution with depth, proportional dry matter partitioning to roots, resource capture coefficients, shoot dry weight at anthesis, specific root weight and water use efficiency. From the current estimates of parameters it is concluded that a larger investment by the crop in fine roots at depth in the soil, and less proliferation of roots in surface layers, would improve yields by accessing extra resources. The economic return on investment in roots for water capture was twice that of the same amount invested for nitrogen capture.

INTRODUCTION

Cereal yields in the UK have increased steadily throughout the latter decades of the 20th century at approx. 0·1 t ha^{-1} y^{-1} through improved cereal varieties and husbandry (Scott et al., 1998). Recently, husbandry improvements have been aimed at improving economic returns and reducing environmental impacts, rather than at just increasing production, such that yield increases have been sustained without increasing the amount of N applied. Confidence to constrain inputs has partly arisen through a new ability to recognize optimum crop canopies (Sylvester-Bradley, 2000), and estimate the returns from 'canopy management' (Sylvester-Bradley et al., 1997b). The major advance in understanding required to achieve this was to consider canopy function as a whole rather than focus on a series of individual physiological attributes. In doing so, the functions of component systems were summarized using key variables (such as the green area index, GAI) which are amenable to management on a field scale and for which quantitative relationships can be described linking their size to inputs, e.g. in the case of GAI, soil and fertilizer N. This approach ignores much physiological detail, especially at cellular levels of organization and below.

Despite these advances in management of the above-ground crop, it has been shown that modern high-yielding varieties may actually be less efficient than their predecessors at capturing soil nitrogen (Foulkes et al., 1998; Sylvester-Bradley et al., 2001). One possible reason for this is that a key part of the plant, the root system, has so far been beyond the reach of management and has received little direct attention for genetic improvement.

The objective of the present work was therefore to develop a theoretical framework that could be applied to the genetic and agronomic improvement of cereal root systems. Specifically, the aim was to model the capture by root systems of two key resources, water and nitrogen, using the minimum amount of crop and soil information necessary to describe their structure and function.

Initially, the study was limited to water and nitrogen because these have the largest and most common effects on crop performance under UK conditions (Archer, 1985; Foulkes et al., 2001). However, yield production goals are not the only motivation to improve the efficiency of resource capture by cereal root systems since there may also be significant environmental benefits. First, sustainable agricultural systems are being extended, and these aim to use soil resources more efficiently. Secondly, climate change may increase the frequency of droughts in temperate cereal-growing regions (Hume and Jenkins, 1998), making access to deep soil water and nutrients more important. In addition, agrochemicals, such as plant growth regulators, and fungicides able to control root diseases (e.g. take all), may be used more effectively if root systems are understood in terms of optimum sizes for economic performance of crops.

This paper reports the development of a model that uses minimal descriptions of the size, distribution and resource capture by a wheat root system to estimate the economic implications of changing key features of that system. Throughout the paper reference is made to the decimal code of Tottman and Broad (1987) to identify the stages of growth during the life cycle of cereals.

MODEL DEVELOPMENT

Rationale

The model was designed for winter wheat (*Triticum aestivum*) grown under typical UK conditions in deep soils of uniform texture and structure. Almost all UK wheat crops receive N fertilizer in order to meet the shortfall between canopy requirement and inherent soil supply. Hence, any additional capture of inherent soil N can be taken to reduce the fertilizer requirement. In the case of water, about 20% of UK crops encounter drought; most often this occurs during grain filling. Hence, any additional capture of water due to improved rooting can be taken to increase yields of these droughted crops. The primary impact of sub-optimal performance of the root system is assumed to be on grain yield, and the model currently only predicts below-ground resource capture over the grain-filling period. The model was fitted to data from the few detailed UK studies on wheat that include data on root system size and distribution with depth in the soil, and N and water uptake during growth (Gregory et al., 1978a, b, 1979a, b; Barraclough, 1984; Barraclough and Leigh, 1984).

Recent studies have shown a uniform distribution of nitrogen with depth at the start of canopy development (Sylvester-Bradley et al., 2001); this becomes skewed to a decline with depth at anthesis following uptake from lower soil layers and replenishment of surface layers through fertilizer addition and mineralization. However, based on the assumption that during canopy expansion nitrogen uptake from topsoils usually takes precedence over deeper uptake, a uniform distribution of nitrogen with depth has been assumed at anthesis here.

In addition, at the start of canopy development when soil is at field capacity, the distribution of soil water with depth is uniform. The development of a soil moisture deficit in the upper layers commences as evapotranspiration starts to exceed rainfall supply. However, this is not usually sufficient to alter significantly the uniformity of water supply with depth until after the time of anthesis in early June, when rainfall supply is no longer sufficient to maintain the upper soil layers, and crops depend more on subsoil resources.

Water and nutrient capture are usually calculated per unit soil volume, using relationships with root length density to quantify the crop's water- and nutrient-absorbing ability per unit volume of soil (Tinker and Nye, 2000). Total uptake is calculated by summing uptake from successive layers of soil within the rooting zone. Thus, information about root distribution with depth is required. Rootable depth defines the total volume of soil available for exploration (Baldwin, 1976; Gales, 1983). In the UK, 2 m is the approximate maximum depth achieved by most winter cereals (Gregory et al., 1978b; Gales, 1983; Lucas et al., 2000), and is generally achieved by the time of anthesis (Fig. 1). Thereafter, senescence can occur at extreme depths and in the upper layers (as in Fig. 1), though the basic pattern of distribution is maintained.

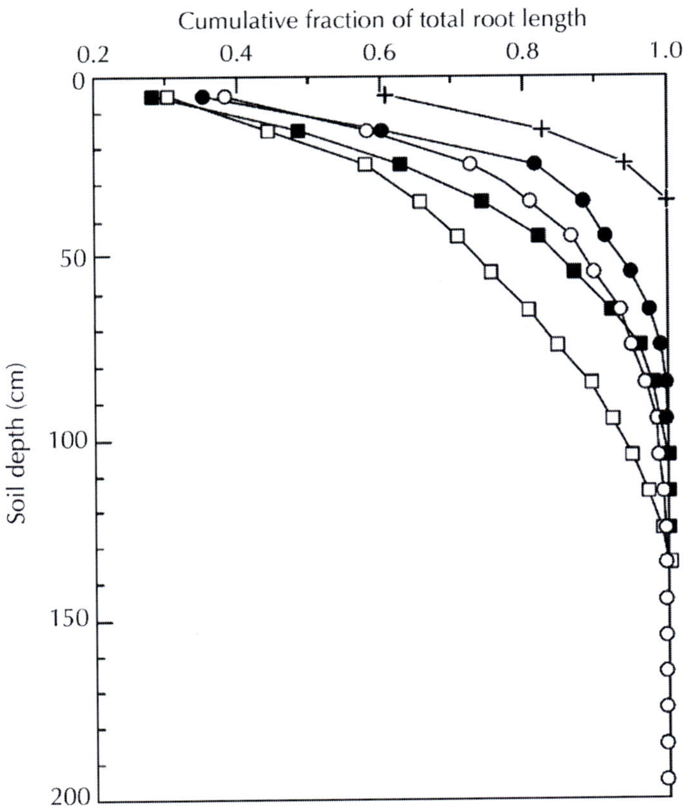

Figure 1: Cumulative fraction of root length with soil depth on five representative dates (in 1975) during the growth of a winter wheat crop ('Huntsman') sown on 30 Oct. 1974 (Gregory et al., 1978a, b). Crosses, 14 January;

closed circles, 8 April (approx. GS31, stem extension); open circles, 10 June (maximum rooting depth achieved); closed squares, 17 June (GS61, anthesis); open squares, 5 August (crop maturity).

The basic morphology of cereal root systems is well known, and because cereal root systems develop in a consistent pattern, they have a relatively predictable architecture (when grown in an homogeneous soil) (Robinson, 1994). Methods are available to describe root development with time and ontogeny, down to the level of individual roots within a hierarchy (Klepper, 1987) and their demo graphy (van Vuuren et al., 1997). However, much of the detailed, dynamic morphology of cereal root systems can be summarized without reducing the resolution of the model significantly.

Similarly, it is possible to reduce much of the complexity of resource capture by crops to a few key variables, and this approach, for example, is routine when modelling light transmission through, and capture by, crop canopies (Pearcy and Valladares, 1999). Water uptake models are often based on radial transport to a single cylindrical root, and the assumption of uniform arrays of such roots in the soil (Gardner, 1960). However, to a reasonable approximation, crop water uptake can also be reduced to an equation relating the amount of water extracted to the mean root length density [$\theta = \theta^{*}\exp(-kL_{v}t)$], where θ is the amount of water extracted, θ^{*} the maximum amount extractable and t is the elapsed time (not Julian date as in Table 1) (Passioura, 1983; Monteith, 1986), k and L_{v} are as defined in Table 1. This has been sufficient to show that mean root length densities (0·5–10 cm cm^{-3}) of cereals in the upper layers of most soils are theoretically adequate to access most of the available soil water within the crop's yield-forming period, and that densities >1 cm cm^{-3} are associated with only small increases in the total amount of water taken up during this period.

Table 1: List of symbols, their definitions and units

Symbol	Definition	Unit
d	Depth in soil profile	cm
D_N	Demand for N during grain filling	kg ha^{-1} d^{-1}
f	Fraction of crop dry matter partitioned to root system	–
G	Grain growth rate	g m^{-2} d^{-1}
	Green leaf area index (GAI)	–
I	Incident photosynthetically active radiation	MJ m^{-2} d^{-1}
k	Resource capture coefficient	cm^2
K	Light extinction coefficient in the canopy	–
L	Total root length	km
L_v	Root length density in soil	cm cm^{-3}
n	Crop growth stage	–
P	Price of grain	£ t^{-1}
R	Total dry weight of roots	g m^{-2}
S	Total dry weight of shoot	g m^{-2}
V_N	Value of nitrogen for grain filling	£ kg^{-1}
V_{water}	Value of water for grain filling	£ mm^{-1}
	Water use efficiency (WUE)	g m^{-2} mm^{-1}
Y	Fraction of the root system accumulated from the soil surface to depth d	–
β	Parameter describing the cumulative distribution of roots with depth	–
ϵ	Radiation use efficiency of the crop	g MJ^{-1}
ϕ	Resource capture as a fraction of the amount available for capture	–
γ	Dry matter fraction of grain	–
σ	Specific root weight	g km^{-1}
t	Julian date	–

With respect to the capture of mobile nutrient resources (water and N), complex models such as those developed by Baldwin et al. (1973) and de Willigen and van Noordwijk (1987) are based on radial flow into dispersed, single roots and then summed to give uptake for the whole root system. These models are probably unnecessarily detailed for our purpose; for example, the problems of radial transport are seldom relevant for crop growth, and can be ignored. The above type of modelling was also carried out by Barber (1995) on plants grown in pots. Sensitivity analyses showed that the length of root was largely unimportant for mobile nutrients (N), but became the most sensitive parameter in the model when considering the uptake of immobile nutrients (P and K). However, being derived from pot-work, the model lacked a term for the relative distribution of roots down a soil profile, which prevented successful translation to field use. One resource capture model for both canopies and root systems that does take into account relative distribution with depth is the WaNuLCAS model of van Noordwijk and Lusiana (1999). Here, tree root length densities in four separate layers down a profile are modelled in relation to their competitive impact on neighbouring crop roots in four spatial zones of an inter-row agroforestry system. It may be that an approach such as ours to the depth distribution of roots would also be appropriate in this model.

Root System Size

The present model is based on the following assumptions:

(1) Adequate water and fertilizer are available during canopy and root system formation, such that crop growth is unrestricted by resource supply before anthesis.

(2) The mass of the root system is predictable from that of the vegetative part of the shoot system throughout the crop's life, and that this partitioning is governed by crop development rather than being a response to the environment.

(3) A mature root system is produced by the beginning of the grain yield forming period [i.e. anthesis, growth stage 61 (GS61), Sylvester-Bradley et al., 1997a], when the main economic impact of the root system is exercised.

(4) The functioning of the root system remains effectively unchanged during grain filling, although the overall size may decline towards harvest.

Typical shoot dry masses of winter wheat in the UK at the start of stem extension (GS31) and at anthesis (GS61) are 0·16 and 1·13 kg m^{-2}, respectively (Sylvester-Bradley et al., 1997a). The root dry mass at

GS61 for the above shoot mass is approx. 146 g m^{-2}, and the maximum root dry weights of UK winter wheat crops are typically between 97 and 170 g m^{-2} (Gregory, 1994).

Total root length (L) is related to root mass (R) by the specific root weight (σ) (Gregory, 1994). Assuming that σ for winter wheat is a constant 4·57 g km^{-1} (as measured by Gregory et al., 1978b), then L at GS61 is 31·9 km m^{-2}. Maximum root lengths of winter wheat crops are between 15 and 32 km m^{-2} (Gregory, 1994).

Root Distribution

It was assumed that:

(1) rooting depth is not restricted by soil physical conditions, nor is it influenced by localized heterogeneities in the soil (cf. Robinson, 1994); and

(2) the distribution of root material with depth is approximated by the Gale and Grigal (1987) equation:

$$Y = 1 - \beta^{d} \quad (1)$$

where Y is the fraction of the root system accumulated from the soil surface to depth d, and β is a parameter that describes the shape of the cumulative distribution with depth. This was first described by Gerwitz and Page (1974) as $Y = 1 - \exp(-\alpha d)$, where $\exp(-\alpha) = \beta$. As β approaches 1, there is a greater proportion of roots deeper in the profile. For $\beta = 0{\cdot}920$, approx. 57% of roots occur in the top 10 cm of the soil profile, whereas this layer contains only about 40% of roots when $\beta = 0{\cdot}950$. Equation (1) is sufficiently robust to summarize root mass and length distributions for a wide range of terrestrial biomes, including temperate grasslands and crops (Jackson et al., 1996, 1997).

β was estimated using the most complete set of data available for the depth distribution of root length (Gregory et al., 1978b; Fig. 1). Figure 2 shows that β increased from 0·940 on 18 March (Julian date 77) to 0·970 by the time the crop reached maturity (5 August; Julian date 217). However, 'parallel curve analysis' (Ross, 1990) using a maximum likelihood program (Ross, 1987) showed that β in Fig. 2 did not change significantly after March. The stability of β during the main growth phase (mean ± s.e. 0·953 ± 0·003) suggests that (in relative terms) root extension may proceed at a similar rate at all depths, and that eqn (1) may be applied before GS61 without altering β. Using the appropriate values of β, *Y* [eqn (1)] and R, root length density (Lv; cm cm^{-3}) in each 10 cm layer of the soil profile can be estimated at each stage of crop growth:

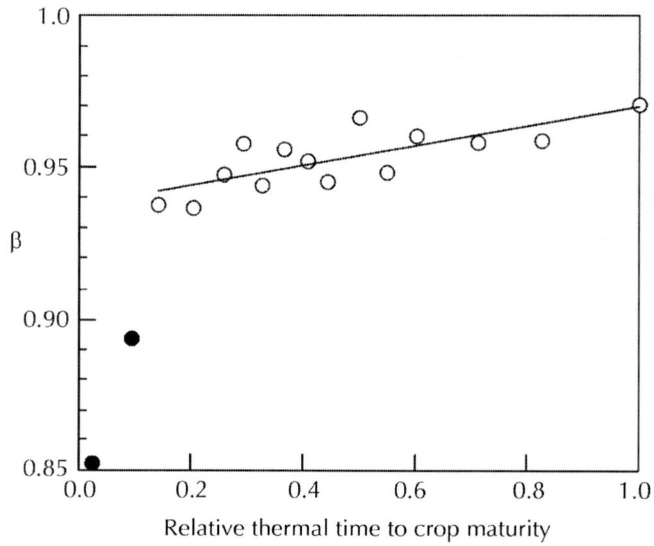

Figure 2: Variation in β (which describes the distribution of root length with soil depth) with relative thermal time (τ_R) for the crop shown in Fig. 1. In this example, τ_R is the number of day-degrees accumulated at each harvest relative to the total accumulated between 1 January and crop maturity (5 August). The linear regression for data between 18 March and 5 August (open circles) is $\beta = 0{\cdot}0313\tau_R + 0{\cdot}938$, $R^2 = 0{\cdot}583$; this period coincided with the major phase of crop growth. Two outlying data points for 14 January and 25 February (closed circles) were omitted from the regression.

$$L_v = (Y_d - Y_d - 10)R/\sigma \qquad (2)$$

Water and Nitrogen Capture

The uptake of water is similar to that of nitrogen (as nitrate) in that, in moist soils used for crop production, both are transported to root surfaces mainly by the mass flow of soil solution (Tinker and Nye, 2000). A generic function was chosen to describe resource capture as a fraction (φ) of that which is potentially available for uptake during the full season:

$$\varphi = 1 - e^{-kL_v} \tag{3}$$

where k is a 'resource capture coefficient' in units of cm^2 [i.e. $1/(cm\ cm^{-3})$]. k subsumes the many details of resource uptake physiology (e.g. the molecular mechanisms of ion and water transport across membranes) and of soil transport (e.g. hydraulic conductivity, ion diffusivity). Larger values of k lead to a more rapid resource depletion for a given root length density. The value of L_v used in eqn (3) can be derived from eqn (2) to calculate φ for each soil layer. Alternatively, a mean L_v for the whole soil profile can be used, as in Fig. 3. To estimate k in the absence of suitable data for wheat, measurements of water and nitrogen use by dryland barley (Gregory and Brown, 1989), which were seasonal averages taken from a water limited system, were used. As such, k is a constant for resource-limited situations, such as those for which this model is intended. A value of $k = 2\ cm^2$ fitted the data for water uptake acceptably well when $Lv > 0{\cdot}6\ cm\ cm^{-3}$, i.e. for all but the most sparse root systems.

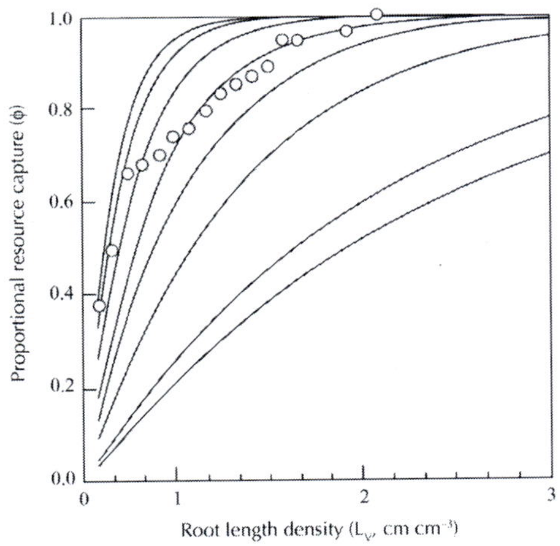

Figure 3: Proportional resource capture (φ), calculated using eqn (3), in relation to root length density (L_v). Each curve is for a different resource capture coefficient (k). These are, from left to right, k = 5, 4, 3, 2, 1·5, 1, 0·5 and 0·4 cm^2, respectively. Data points are for water capture by barley crops relative to total amount available in the soil, and for root length densities of those crops measured at anthesis (Gregory and Brown, 1989). The data fit reasonably well to the $k = 2\ cm^2$ line when $L_v > 0{\cdot}6\ cm\ cm^{-3}$. N capture would be expected to follow a similar trajectory. The curve for $k = 0{\cdot}4\ cm^2$ gives $\varphi \approx 1$ at $L_v \approx 10\ cm\ cm^{-3}$ as would be expected theoretically for P capture.

Resource capture by the whole root system is the accumulated capture by roots in each soil layer. Assuming a maximum possible rooting depth (e.g. 2 m for UK winter cereals: Gregory et al., 1978b; Gales, 1983; Lucas et al., 2000), and a uniform distribution of available resources with depth, total capture is easily calculated as mean φ averaged over all soil layers, including those containing no roots. On this basis, a winter wheat crop could capture about 55 % of available water and N by anthesis, which is equivalent to all the resources contained within the top 1·1 m of the soil profile.

Economic Evaluation

We consider the case where drought limits grain filling by reducing water capture compared with a well-watered crop, whilst N is not limiting. In this assessment it is assumed that all dry matter accumulation in the shoot post-anthesis is diverted to grain filling. The economic value (V_{water}) of water use during grain filling is:

$$V_{water} = WP/\gamma \qquad (4)$$

where W is WUE, P is the price per tonne of grain and γ is the dry matter fraction at which the grain is sold. A typical recent price of wheat in the UK is £70 t^{-1} (P), and grain is sold at 85 % dry matter (γ). For a WUE of 5 g m_{-2} mm^{-1} (Foulkes et al., 2001), Vwater was calculated by eqn (4) to be £4·12 mm^{-1} ha^{-1}.

In N-limiting conditions whilst soil water is abundant, it is assumed that extra N capture during grain filling delays the onset of senescence and the end of the grain-fill period. Grain growth rate (G) during grain fill can be calculated from eqn (5) (Sylvester-Bradley et al., 1997b) where all growth is directed towards grain fill rather than vegetative biomass.

$$G = \epsilon/(1 - e^{-KA}) \qquad (5)$$

where A is GAI ϵ, I and k are as defined in Table 1. For typical values of $\epsilon = 2{\cdot}4$ g MJ^{-1}, $I = 9$ MJ m^{-2} d^{-1}, $K = 0{\cdot}45$ and GAI = 6, the grain growth rate, G, during filling is about 200 kg ha^{-1} d^{-1}.

An equation analogous to eqn (4) was used to calculate the economic value (V_N) of N used in grain filling:

$$V_N = GP/\gamma D_N \qquad (6)$$

where D_N is the daily demand for N during grain filling. If 200 kg N ha^{-1} [eqn (5)] is required over a typical grain-filling period lasting 40 d, DN is about 5 kg ha^{-1} d^{-1}, which must be translocated from the canopy. Assuming a grain price of £70 t^{-1}, then the value of extra N during grain filling amounts to £3·32 kg^{-1} N.

The sensitivity of the net economic return to the values of the main variables was tested by varying the value of each over a defined range whilst holding the values of the others constant. These ranges were: β, 0·955–0·985; σ, 2·5–7·5 g km^{-1}; k, 1–3 cm^2; and f, 0·05–0·2. The effects on water uptake of varying shoot dry weight at anthesis (SGS61, 0·8–1·4 kg m^{-2}) and WUE (4–6 g m^{-2} mm^{-1}) were also tested, as were variations in grain growth rate (G, 0·15–0·25 t ha^{-1} d^{-1}) and grain demand for N (DN, 3–5 kg $ha^{-1}d^{-1}$) for their effects on N uptake.

The parameters β, WUE and σ had the largest effects on economic yield in relation to water uptake (Fig. 4) and, in relation to *N* uptake, β and σ were again important (Fig. 5). The effects of all variables on economic yield in relation to N uptake were less than half those in relation to water uptake.

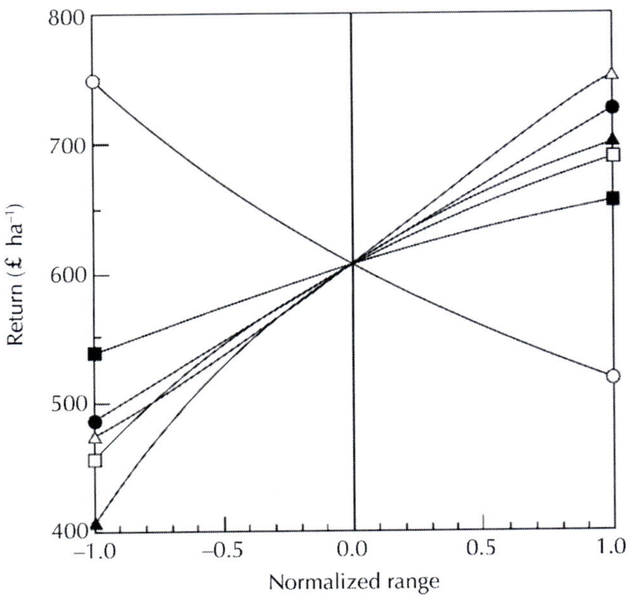

Figure 4: The sensitivity of the economic return to variations in water capture during grain filling, as affected by individual variables in the model. The

value of each variable was adjusted singly over a range normalized to zero at its mid-point. Open triangles, root distribution with soil depth (β, 0·955–0·985); closed triangles, dry matter partitioning to roots (f, 0·05–0·2); open squares, resource capture coefficient (k, 1–3 cm^2); closed squares, shoot dry weight at anthesis (S61, 0·8–1·4 kg m^{-2}); open circles, specific root weight (σ, 2·5–7·5 g km^{-1}); closed circles, water use efficiency (WUE, 4–6 g m_{-2} mm^{-1}).

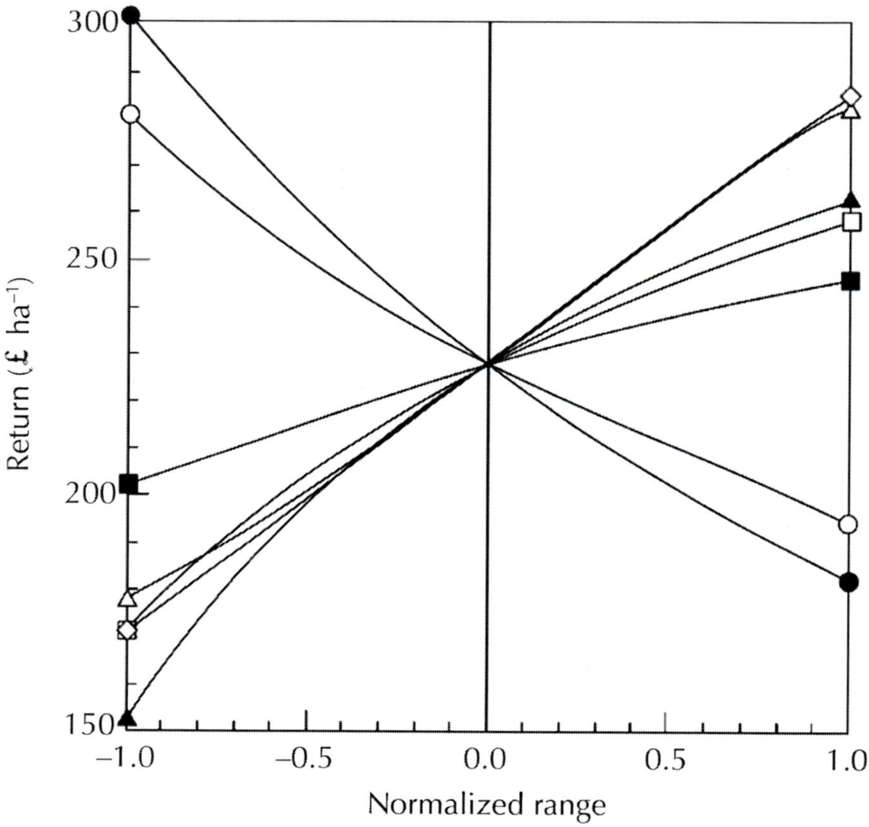

Figure 5: Sensitivity of the economic return to variations in N capture during grain filling, as affected by individual variables in the model. The value of each variable was adjusted singly over a range normalized to zero at its mid-point. Symbols and ranges as for Fig. 4, with the exception of: closed circles, grain demand for N (DN, 3–5 kg $ha^{-1}d^{-1}$); open diamond, grain growth rate (G, 0·15–0·25 t ha^{-1} d^{-1}).

DISCUSSION

Outputs from the Model and their Significance

The theoretical framework that is developed here allows:

(1) the size and distribution of winter wheat root systems at anthesis to be described;

(2) water and nitrogen capture, as a fraction of the amounts available for capture, to be estimated, along with the use of these resources in yield formation; and

(3) the economic effects of changes in single root system characters on water and nitrogen capture to be evaluated.

According to the model, the vertical distribution of the root system, as described by β, can have a profound effect on yield. Larger values of β give a greater economic return for water and *N* capture. This implies that a more uniform distribution of roots with depth could, in theory, improve the economic return from a winter wheat crop. This is because such crops could more readily access resources deeper in the soil that would be unavailable to crops with smaller β values. If the same root mass is distributed more evenly with depth, roots in upper layers will, on average, be less concentrated. This will delay the point at which inter-root competition begins in these layers and, therefore, prolong the effectiveness of each of the roots in resource capture. At the same time, deeper layers would more quickly achieve root length densities adequate for maximum resource capture. The effectiveness of the whole root system is then increased, compared with that of a less evenly distributed root system. However, it should be noted that this enhanced effectiveness would be restricted if the resource also declined with depth, as N may do in many cases (Sylvester-Bradley et al., 2001). Conversely, it may become even more important for resource distributions which increase with depth, as may be the case for water in drought years.

The partitioning (f) of dry matter between shoot and root systems was also influential on grain yield, as was the specific root weight (σ),which is approximately proportional to the square of the mean radius. The model predicts greater economic returns as a result of water and N uptake from crops with greater partitioning of dry matter to roots (larger f). This prediction agrees with one made by Baldwin (1976) who modelled total crop yields on the basis of a detailed treatment of soil transport, water and nutrient uptake, and root–shoot partitioning. However, our model also predicted maximum returns from crops with finer roots (smaller σ), contradicting Baldwin's (1976) conclusion: he predicted a 32 % increase in yield for a doubling of mean root radius (roughly equivalent to doubling $\sigma 0{\cdot}5$). Rather than the extra dry matter being invested in thicker roots to increase resource translocation (Baldwin, 1976), the model presented here suggests that it would be better invested in expanding the absorptive area (per unit mass) at lower depths in the soil to increase overall resource capture.

The Implications of Model Outputs for the Manipulation Wheat Root Systems

If the location of appropriate characters can be identified in gene sequences then genotypes with more efficient root systems may be identified, bred or engineered. However, our model takes no account of possible genetic linkages or interactions between characters, and these could confound attempts to genetically manipulate one character independently of others. As cereal genetic maps become increasingly detailed and include root traits, the potential for targeted trait manipulation can be assessed a priori. For example, root dry weight in barley (Hordeum vulgare) is mapped close to the locus of a major dwarfing gene (ari-eGP) with widespread developmental effects (R. P. Ellis pers. comm.). Attempts to manipulate root dry weight in that species could possibly be confounded by simultaneous and undesirable changes in other important characters. Whether the same constraint would apply to wheat would depend on the degree of synteny between the species (Freeling, 2001), but this remains to be determined.

Characters may also be constrained allometrically, or may vary with the environment, limiting the extent to which each could be manipulated independently, e.g. via plant growth regulators. For example, an increase in partitioning of dry matter to roots (f) is predicted to increase the economic return. Yet, logically, crop growth rate will begin to fall if f becomes too large and the canopy too small for maximum photosynthesis. It might prove impossible to simultaneously increase f and shoot weight at anthesis to maximize the economic return, at least beyond certain (as yet unknown) limits.

The resource capture coefficient (k) proved a less important influence on the economic return from water and N capture because the range of root length densities chosen to test its impact on resource capture (Fig. 3) was relatively small. Much greater variation has been reported for root systems in general, so it will be important to validate our assumption of smaller variation in commercial wheat crops within the UK.

One of the advantages of the present model is that it needs crop information for only one growth stage (GS61), and the outputs from the model were achieved using minimal information about the crop and soil. In this respect, the model offers considerable practical utility: decisions in crop husbandry must necessarily be based on a few simple observations. The model predicts that at current costs, water is a more valuable resource than N in terms of the probable returns from managing the root system, but this will depend on circumstances; clearly, water will be the more important resource on light-textured soils, whereas N will be more important on deep, retentive soils. Particularly in the case of N, it may be useful to extend the use of the model to assess effects at other growth stages, to assist earlier decision making.

Future Developments

The present model was based on several assumptions or 'typical' values. Some of these assumptions (e.g. a maximum rooting depth of 2 m) were derived, for expediency, from a limited set of experiments, and most of these assumptions need testing for a wider range of soil types and geographical locations. For example, it is extremely unlikely that physical impedances will not limit rooting depth in certain soils, e.g. shallow soils overlying rock, or waterlogged soils. The effects of such limitations on the capture of different resources are difficult to predict. The distribution of wheat in relation to climate and soil type must also be taken into account, as must the resource limitations that may be most important economically in different regions. In reviewing various constraints on yield in the UK, Gales (1983) noted that crop root systems could extract water to depths of 1·5–2·0 m on sandy loams and clay soils. Although the amount of water considered 'available' was that held between 0·005–1·5 MPa tension, abstraction below 0·5 m tended to be only of 'easily available water' held at tensions between 0·005–0·2 MPa. Taking this into account, Gale concluded that drought in the UK would be most likely to occur in crops on clay soils (i.e. with less easily available subsoil water), even though such soils contain ample total amounts of water. Many soils have an uneven distribution of resources owing to fertilizer applications and the periodic saturation of topsoil with rainwater. Our assumption that water and N are distributed evenly throughout the whole rooting volume is an obvious oversimplification. However, it would be easy to relax this assumption and impose an initially non-uniform distribution if sufficient information were available. The effects of uneven initial distributions on the model's predictions can be tested using assumed patterns for water and N analogous to eqn (3), though a potential difficulty lies in defining the amounts available from rainfall and mineralization during crop growth. Inclusion of transport between soil layers would be more difficult, though simple algorithms exist to achieve this (e.g. Campbell, 1985).

A related issue concerns the effects of resource availability during the formative growth phases on and k during yield formation. For instance, irrigation and N fertilizer can cause root proliferation in surface soil (Barraclough et al., 1989; Robinson, 1994), so the distribution of water availability earlier in the crop's growth may alter at anthesis. It is possible that k also varies under drought or with N supply. This possibility will require testing, as will the determination of k as a function of root length density for different nutrients, but without the confounding effects of soil depth. Finally, the logical goal of developing such a framework is to consider whether root systems are actually optimal for resource capture. To do this, it will be necessary to calculate not only their ability to capture resources, but also the costs of resource capture, possibly in terms of carbon (cf. Robinson, 2001). The present model indicates a possible over-abundance of roots in the surface layers for water and N capture (though not necessarily for P) during later growth stages: it will be useful to test genotypes with larger values.

Another obvious aspect of the model that requires further development is to consider P capture. Soil resources other than water and N may have only weak effects on wheat yield in most fields in the UK owing to regular fertilizer use (Archer, 1985). Nevertheless, it will be necessary to check that effects of changes in root system characters on P capture do not compromise the conclusions drawn for water and N. In most soils, equilibrium concentrations of P in soil solution, and diffusivities of P through the soil, are limited by buffering reactions (Tinker and Nye, 2000) and, consequently, the mean root length densities required to theoretically access all of the available P are about an order of magnitude larger than those needed to capture water and N (van Noordwijk, 1983). In Fig. 3, relative P uptake would probably follow trajectories described by $k < 0{\cdot}5$ and, in theory, P uptake should be more closely correlated with root length density than are the uptake of water and N. However, this is not always borne out by experiments (Otani and Ae, 1996), and further investigation is required. P uptake will require a different conceptual approach from that used for N and water uptake because P concentrations in solution are low and absorption processes will have to be included. Rather than being based on total root length, P uptake may depend more on new root growth (Yanai, 1994). At a later stage, questions about K, S and micronutrients may also need to be addressed.

CONCLUSIONS

There remains considerable scope for exploring further simplifications in nutrient uptake models with the aim of improving their practicality. The approach adopted here has been guided by this aim. By constructing a summarizing framework of root resource capture, we have shown that it is possible to evaluate the value of the size and distribution of a wheat root system, and its water- and N-capturing functions, in terms of the crop's final economic performance. This is a major step forward and opens the way to assessing the efficiency of different root systems and, furthermore, to deriving an optimal root system, or ideotype (cf. Jones, 1983, p. 274), for UK conditions.

Acknowledgement

This work was funded by the Department for Environment, Food & Rural Affairs (Project AR 0902).

REFERENCES

1. AddiscottTM, Bailey NJ, Bland, GJ, WhitmoreAP.1991. Simulation of nitrogen in soil and winter wheat crops: a management model that makes the best use of limited information. Fertilizer Research27: 305–312.
2. ArcherJ.1985. Crop nutrition and fertiliser use. Farming Press. BaldwinJP.1975. A quantitative analysis of the factors affecting plant nutrient uptake from some soils. Journal of Soil Science26: 195–206.
3. BaldwinJP.1976. Competition for plant nutrients in soil; a theoretical approach. Journal of Agricultural Science87: 341–356.
4. BaldwinJP, Nye PH, Tinker PB.1973. Uptake of solutes by multiple root systems from soil. III. A model for calculating the solute uptake by a randomly dispersed root system developing in a finite volume of soil. Plant and Soil38: 621–635.
5. BarberSA.1995. Soil nutrient bioavailability. A mechanistic approach. New York: John Wiley & Sons, Inc.
6. BarracloughPB.1984. The growth and activity of winter wheat roots in the field: Root growth of high-yielding crops in relation to shoot growth. Journal of Agricultural Science103: 439–442.
7. BarracloughPB. 1986. The growth and activity of winter wheat roots in the field: nutrient inflows of high yielding crops. Journal of Agricultural Science106: 53–59.
8. BarracloughPB, Leigh RA.1984. The growth and activity of winter wheat roots in the field: the effect of sowing date and soil type on root growth of high yielding crops. Journal of Agricultural Science103: 59–74.
9. BarracloughPB, Kuhlmann H, Weir AH.1989. The effects of prolonged drought and nitrogen fertilizer on root and shoot growth and water uptake by winter wheat. Journal of Agronomy and Crop Science163: 352–360.
10. CampbellGS1985. Soil physics with BASIC. Transport models for soil-plant systems. Amsterdam: Elsevier.

11. de WilligenP, van Noordwijk M.1987. Roots, plant production and nutrient use efficiency. PhD Thesis, Wageningen Agricultural University, Wageningen.
12. de WilligenP, Nielsen NE, Claassen N, Castrignanò AM.2000. Modelling water and nutrient uptake. In: Smit AL, Bengough AG, Engels C, van Noordwijk M, Pellerin S, van de Geijn SC, eds. Root methods: a handbook. Berlin: Springer, 509–543.
13. FoulkesMJ, Scott RK, Sylvester-Bradley R.2001. A comparison of the ability of wheat cultivars to withstand drought in UK conditions: resource capture. Journal of Agricultural Science137: 1–16.
14. FoulkesMJ, Sylvester-Bradley R, Scott RK.1998. Evidence for differences between winter wheat cultivars in acquisition of soil mineral nitrogen and uptake and utilization of applied fertilizer nitrogen. Journal of Agricultural Science130: 29–44.
15. FreelingM.2001. Grasses as a single genetic system. Reassessment 2001. Plant Physiology125: 1191–1197.
16. GaleMR, Grigal DF. 1987. Vertical root distributions of northern tree species in elation to successional status. Canadian Journal of Forest Research17: 829–834.
17. GalesK.1983. Yield variation of wheat and barley in Britain in relation to crop growth and soil conditions – a review. Journal of the Science of Food and Agriculture 34: 1085–1104.
18. GardnerWR.1960. Dynamic aspects of water availability to plants. Soil Science89: 63–73.
19. GerwitzA,Page ER. 1974. An empirical mathematical model to describe plant root systems. Journal of Applied Ecology11: 773–781.
20. GregoryPJ.1994. Resource capture by root networks. In: Monteith JL, Scott RK, Unsworth MH, eds. Resource capture by crops. Nottingham: Nottingham University Press, 77–97.
21. GregoryPJ, Brown SC.1989. Root growth, water use and yield of crops in dry environments: what characteristics are desirable? Aspects of Applied Biology22: 235–243.
22. GregoryPJ, Crawford DV, McGowan M.1979a. Nutrient relations of winter wheat. 1. Accumulation and distribution of Na, K, Ca, Mg, P, S and N. Journal of Agricultural Science93: 485–494.

23. GregoryPJ, Crawford DV, McGowan M.1979b. Nutrient relations of winter wheat. 2. Movements of nutrients to the root and their uptake. Journal of Agricultural Science93: 495–504.
24. GregoryPJ, McGowan M, Biscoe PV.1978a. Water relations of winter wheat. 2. Soil water relations. Journal of Agricultural Science91: 103–116.
25. GregoryPJ, McGowan M, Biscoe PV, Hunter B.1978b. Water relations of winter wheat. 1. Growth of the root system. Journal of Agricultural Science91: 91–102.
26. HumeM, Jenkins G.1998. Climate change scenarios for the United Kingdom, London: UKCIP Technical Report No. 1.
27. JacksonRB, Mooney HA, Schulze ED.1997. A global budget for fine root biomass, surface area, and nutrient contents.Proceedings of the National Academy of Sciences of the USA94: 7362–7366.
28. JacksonRB, Canadell J, Ehleringer JR, Mooney HA, Sala OE, Schulze ED.1996. A global analysis of root distributions for terrestrial biomes. Oecologia108: 389–411.
29. JonesHG.1983. Plants and microclimate. Cambridge: Cambridge University Press.
30. KlepperB.1987. Origin, branching and distribution of root systems. In: Gregory PJ, Lake JV, Rose DA, eds. Root development and function. Cambridge: Cambridge University Press, 103–124.
31. LucasME, Hoad SP, Russell G, Bingham IJ.2000. Management of cereal root systems. HGCA Research Review 43. London: Home Grown Cereals Authority.
32. MonteithJL.1986. How do crops manipulate water supply and demand? Philosophical Transactions of the Royal Society of London A316: 245–259.
33. OtaniT, Ae N.1996. Sensitivity of phosphorus uptake to changes in root length and soil volume. Agronomy Journal88: 371–375.
34. PassiouraJB. 1983. Roots and drought resistance. Agricultural Water Management7: 265–280.
35. PearcyRW, Valladares F.1999. Resource capture by plants: the role of crown architecture. In: Press MC, Scholes J, Barker MG, eds. Physiological plant ecology. Oxford: Blackwell Science, 45–66.

36. RobinsonD.1994. The responses of plants to non-uniform supplies of nutrients. New Phytologist127: 635–674.
37. RobinsonD.2001. Root proliferation, nitrate inflow and their carbon costs during nitrogen capture by competing plants in patchy soil. Plant and Soil232: 41–50.
38. RossGJS.1987. Maximum likelihood program, version 3.08. Oxford: Numerical Algorithms Group.
39. RossGJS.1990. Nonlinear estimation. New York: Springer.
40. ScottRK, Sylvester-Bradley R.1998. Crop physiology, assessment and management. Home Grown Cereals Authority, 1998 R & D Conference 2.1–2.14.
41. Sylvester-BradleyR, Scott RK, Clare RW.1997a. The wheat growth guide. London: Home Grown Cereals Authority.
42. Sylvester-BradleyR, Scott RK, Stokes DT, Clare RW.1997b. The significance of crop canopies for N nutrition. Aspects of Applied Biology50: 103–116.
43. Sylvester-BradleyR, Stokes DT, Scott RK.2001. Dynamics of nitrogen capture without fertilizer: the baseline for fertilizing winter wheat in the UK. Journal of Agricultural Science136: 15–33.
44. Sylvester-BradleyR, Spink J, Foulkes MJ, Bryson RJ, Scott RK, Stokes DT, King JA, Parish D, Paveley ND, Clare RW.2000. Sector Challenge Project – Canopy Management in practice. Home Grown Cereals Authority, 2000 R & D Conference, 11.1–11.14.
45. TinkerPB, Nye PH.2000. Solute movement in the rhizosphere. Oxford: Oxford University Press.
46. TottmanDK, Broad H.1987. The decimal code for the growth stages of cereals with illustrations. Annals of Applied Biology110: 441–454.
47. van NoordwijkM.1983. Functional interpretation of root densities in the field for nutrient and water uptake. In: Böhm W, Kutschera L, Lichtenegger E, eds. Wurzelokologie und ihre Nutzanwendung. Irdning: Bundesanstalt für alpenlandische Landwirtschaft, 207–226.
48. van NoordwijkM, Lusiana B.1999. WaNuLCAS, a model of water, nutrient and light capture in agroforestry systems. Agroforestry Systems43: 217–242.

49. vanVuurenMMI, Robinson D, Fitter AH, Chasalow SD, Williamson L, Raven JA.1997. Effects of elevated atmospheric CO2 and soil water availability on root biomass, root length, and N, P and K uptake by wheat. New Phytologist 135: 455–465

50. YanaiRD.1994. A steady-state model of nutrient uptake accounting for newly grown roots. Soil Science Society of America Journal58: 1562–1571.

Chapter 4

Perennial Forages as Second Generation Bioenergy Crops

Matt A. Sanderson and Paul R. Adler

USDA-ARS, Pasture Systems and Watershed Management Research Unit, Building 3702, Curtin Road, University Park, PA 16802-3702

ABSTRACT

The lignocellulose in forage crops represents a second generation of biomass feedstock for conversion into energy-related end products. Some of the most extensively studied species for cellulosic feedstock production include forages such as switchgrass (*Panicum virgatum* L.), reed canarygrass (*Phalaris arundinacea* L.), and alfalfa (*Medicago sativa* L.). An advantage of using forages as bioenergy crops is that farmers are familiar with their management and already have the capacity to grow, harvest, store, and transport them. Forage crops offer additional flexibility in management because they can be used for biomass or forage and the land can be returned to other uses or put into crop rotation.

Estimates indicate about 22.3 million ha of cropland, idle cropland, and cropland pasture will be needed for biomass production in 2030. Converting these lands to large scale cellulosic energy farming could push the traditional forage-livestock industry to ever more marginal lands. Furthermore, encouraging bioenergy production from marginal lands could directly compete with forage-livestock production.

INTRODUCTION

Second generation bioenergy crops [1], based on perennial forage crop species, are considered to be the future of the bioenergy industry and are the focus of intense research [2-4]. Compared with the first generation biofuels based on annual grain crops, perennial biomass crops require fewer inputs, produce more energy, and reduce greenhouse gas (GHG) emissions more than annual cropping systems [5].

Switchgrass (*Panicum virgatum* L.) is particularly compelling in North America because of its relatively low production inputs and costs, perennial growth habit, and adaptability to a broad range of growing conditions [6]. Perennial forage crops such as alfalfa (*Medicago sativa* L.), reed canarygrass (*Phalaris arundinacea* L.), napiergrass (*Pennisetum purpureum Schumach.*), and bermudagrass (*Cynodon* spp.) among many others, could also serve as perennial bioenergy crops for specific agroecoregions of North America [7].

These second generation bioenergy crops historically have been used for grazing and forage and were the original energy feedstocks for draft animal power [8]. Perennial forage crops currently supply the energy that fuels approximately 100 million ruminant animals on USA farms with a total estimated economic value of $US 39 billion [9]. The lignocellulose in perennial forage crops represents a vast and renewable source of biomass feedstock for conversion into the second generation of biobased products [10]. Indeed, the concepts of "fuel farming" are not new [e.g., 11-14]; however with new technologies and processes for biomass production and conversion nearing commercial reality [15], perennial forage crops could once again fuel agriculture.

Increasingly, farmers must consider managing for multifunctionality and include effects on environmental quality in their management decision-making [16]. Thus, grassland producers will need to consider managing for emerging ecosystem services such as enhancement of carbon (C) sequestration, mitigation of GHG emissions, and to capitalize on new opportunities, such as bioenergy production, to diversify the forage-livestock system to achieve these outcomes in the future [17]. In this paper we discuss some of the forage crops proposed for bioenergy use along with their environmental impacts, energy balance, research needs, and explore the implications of their widespread use.

PERENNIAL FORAGE SPECIES FOR BIOENERGY

Of the many species of perennial forage crops available [18] only a few have been researched intensively for biomass. We briefly describe some of those species.

Switchgrass

In the USA, research on perennial bioenergy crops during the last two decades has focused on switchgrass, a C_4 native warm-season perennial grass [4, 19]. Attributes of switchgrass desirable for bioenergy cropping include its demonstrated high productivity across many environments, suitability for marginal and erosive land, relatively low water and nutrient requirements, and positive environmental benefits [6, 20]. As potential disadvantages, switchgrass can be slow to establish, and productive stands often take two years to develop. 'Alamo' is a well adapted cultivar for the southern USA, and 'Cave-in-Rock' is a broadly adapted cultivar suitable for the mid-Atlantic, Northeast, and Midwest regions of the USA (Tables 1 and 2) [19, 20]. Newer varieties of switchgrass with improved biomass yield and chemical composition have been released [22-24]. Compared with other crop species, switchgrass has received relatively little plant breeding attention and most cultivars of this species are not far removed from native germplasm [20]. In some cases, cultivars of switchgrass could not be distinguished genetically from natural populations [25]. Thus, there is tremendous genetic variability and great potential for germplasm improvement [25-27].

Miscanthus

Miscanthus, a C_4 grass native to Asia, is viewed as a model herbaceous biomass feedstock for Europe [28]. Although not used as a forage crop because of its morphology, *Miscanthus* is used as an ornamental plant in the USA. Researched extensively from northern to southern Europe (Table 1), the primary use of *Miscanthus* biomass is envisioned principally as a feedstock in combustion steam generating electrical plants [28]. *Miscanthus* has good cold tolerance for a C_4 species and is winter hardy in temperate regions of Europe [29]. *Miscanthus* has a low requirement for N fertilizer because it efficiently recycles N between aboveground biomass and storage structures (rhizomes) belowground [30]. In most of Europe, *Miscanthus* grows from April until November; however, harvest of the previous year's biomass in February or March is recommended because moisture content and alkali elements in the standing biomass are reduced [31]. A drawback to *Miscanthus* use is that it must be established and propagated vegetatively via rhizome cuttings, which delays full production until the second or third year and also requires irrigation and energy inputs during greenhouse propagation [32].

A virtual comparison of *Miscanthus* production in Europe with switchgrass production in North America speculated that *Miscanthus* could produce twice as much biomass as switchgrass [33]. *Miscanthus* yielded 33% more biomass (18.1 vs. 14.1 Mg ha-1) than 'Kanlow' switchgrass grown on a heavy clay soil in southwestern Germany [34]. The authors cautioned, however, that Kanlow switchgrass may not have been well adapted to the soils and environment of the experimental site in Germany.

Reed Canarygrass

Scandanavian research on bioenergy crops has identified reed canarygrass as a highly productive perennial grass for northern Europe [28, 31]. Reed canarygrass is a C_3 grass commonly used for hay and grazing that is well adapted to temperate agroecoregions and does well on wet soils [35]. Yields of reed canarygrass in Indiana, USA averaged 10 Mg ha^{-1} under low or high N fertilizer management (Table 1) [36]. Field trials in Iowa USA demonstrated biomass yields of 8.6 Mg ha^{-1} with 140 kg N ha^{-1} at two locations over 5 yr (Table 1) [37]. Some evidence exists for an efficient internal N recycling mechanism from shoots to roots in reed canarygrass [38]. Similar to switchgrass, reed canarygrass can be slow to establish, and yields are low in the seeding year. Reed canarygrass, however, can be an invasive species in native wetlands [39].

Table 1: Example biomass yields from selected perennial crops and lands

Crop	Location and description	N rate kg ha-	Biomass yield		Source
			Range Mean Mg	Mean ha^{-1}	
Switehgrass	Field-scale plots (3 to 9.5 ha) on 10 farms in Nebraska, South Dakota, and North Dakota USA harvested for 5 yr	0-212	5.2-11.1		49
Miscanthus	Experimental plots harvested for 3 yr in Denmark	60	1.4-18.2	9.1	106
Miscanthus	Experimental plots irrigated and harvested for 3 yr in Portugal	60	7.5-40.9	25.2	106
Reed canarygrass	xExperimental plots in Indiana USA harvested for 3 yr	0-168	9.4-10.1	10.0	36
Reed canarygrass	Experimental plots at two sites in Iowa USA harvested for 5 yr	140	5.5-102	7.7	37

Alfalfa	Experimental plots at two sites in Minnesota USA harvested for 2 yr	0	7.0-12.0		42
Bermudagrass	Three experimental plot sites in Georgia USA for 3 yr	NR[1]	12.8-	15.0	45
			19.9		
Napicrgrass	Experimental plots in northern Florida USA harvested for 2 yr	200		46.3	107
Eastern gamagrass	Summary of studies from nine states in the eastern USA	84-301	6.5-15.9		47
Prairie cordgrass	Experimental plots in South Dakota USA harvested for 4 yr	0	4.6-8.6	6.4	48
Pasture on marginal land	10 pasture sites in southern Iowa USA	NR	0.8-8.2	4.2	58
CRP land	34 sites in seven northeastern USA states	0		6.6	61
CRP land	Experimental plots at three South Dakota USAsites harvested for 3 yr	0-224	2.56.0		62
Low-input high-diversity prairie	I to 16 plant species grown in small plots grown for 10 yr at Cedar	0		3.7	65
	Creek, Minnesota USA				

Table 2: Example biomass yields from several switchgrass cultivars in the USA, Canada, and Europe

	North	South			Midwest	Pennsylvania		Mid-Atlantic states7			Alabama9		North			
Cultivar	Dakota1	Dakota2	Wisconsin2	Iowa3	States4	A5	B6	1-cut	2-cut	Texas8	A	B	Carolina10	Quebec11	Greece12	Italy12
	Mg dry biomass ha^{-1}															
Cave-in-Rock	4.9	3.8	14.3	9.3	9.2	8.6	8.2	10.8	15.4	2.6			12.4	12.2	12.5	7.7
Dacotah	5.4	2.9	7.4													
Forestburg		3.9	9.4	6.9												
Shawnee	5.6	5.1	11.4	8.8		8.5										
Sunburst	7.4	4.6	11.5	6.8	8.8									10.6		
Trailblazer	6.9	4.6	11.0	7.9		6.7	12.4									
Alamo				12.1				15.2	16.3	15.3	23.0	12.9	14.2			
Pathfinder					8.3		11.0							11.5		
Shelter								10.3	13.6							
Kanlow							12.1	15.0	16.4	11.0	18.2	11.6			17.1	10.0
Blackwell					9.1											
NJ50							12.6									
Summer	5.5														14.6	7.4
BoMaster													15.8			
Performer													12.8			

Alfalfa

Alfalfa is one of the world's oldest forage crops and is perhaps the highest value forage in North America [40]. Researchers in the 1990s envisioned alfalfa as a dual-use crop to simultaneously supply both biomass feedstock and a high quality animal feed [41]. In the dual-use system, leaves are separated for high-value, high-protein feed (an additional income stream) and the fibrous, lignified stems are combusted in an integrated gasification combined-cycle system to produce electricity. The proposed system recommended a two-cut harvest management to optimize economics, yield of stem and leaf, and wildlife habitat. Genetic selection efforts concentrated on lines developed for stiff stems with increased internode length to be grown under infrequent harvest (Table 1) [42]. Experimental biomass-type alfalfa germplasm has greater stem cell wall polysaccharide concentrations along with greater stem lignin concentrations, which contributed to greater stem dry matter yield and theoretical ethanol yields compared with hay-type alfalfas [43].

Other Species

Several other subtropical and tropical grasses have been evaluated as biomass crops in the southern region of the USA. The long, warm growing season and high rainfall in the southeastern region provide conditions for high yields ranging from 10 to 40 Mg ha^{-1} dry matter [44]. Bermudagrass (*Cynodon* spp.) is grown as a hay and grazing crop on about 4 million ha of the southern USA.

Bermudagrass biomass yields have ranged from 13 to 20 Mg ha^{-1} at several locations in Georgia (Table 1) [45]. Napiergrass is a tall growing, perennial tropical grass with yields of up to 30 Mg ha^{-1} in Florida [46]. Eastern gamagrass (*Tripsacum dactyloides* L.) and prairie cordgrass (*Spartina pectinata* Link.) have also been explored as potential perennial grass feedstocks [47, 48].

There are many crops that could potentially be used to supply biomass feedstock to the emerging cellulosic (second generation) energy industry. This industry will require multiple sources of biomass and multiple biomass feedstocks for specific agroecoregions. The crops detailed here will probably be the first among many feedstocks to come from perennial forages.

MANAGEMENT FOR BIOENERGY CROPPING

Farmers are familiar with the agronomic management of perennial forage crops. Further, some of the machinery, technology, and infrastructure needed to plant, harvest, store, and transport forage crops can be used in bioenergy production. Details regarding specific management practices for bioenergy as they relate to forage production are discussed elsewhere [6, 7]. Principal management factors that influence biomass productivity and feedstock quality of a particular species include (i) rapid seedling establishment to reduce the time to productive stands, (ii) optimizing fertilizer inputs, and (iii) harvest management to optimize yield, persistence, and feedstock quality.

Establishment is a critical phase in forage and bioenergy crop production. Establishment failures substantially reduced switchgrass biomass and net energy yields in on-farm research [49]. Difficulties in stand establishment are often related to poor seed quality, improper planting depth, poor seedbed preparation, lack of weed control, and variable soil and weather conditions.

Fertilizer inputs must be optimized in biomass cropping systems because significant fossil fuel energy is used in the synthesis of the fertilizer. On-farm research with switchgrass production in the central Great Plains of the USA documented that fertilizer N accounted for 67% of the fossil fuel energy inputs to the biomass production system [49]. Inputs of N fertilizer also can be used as an indicator of potential environmental consequences of bioenergy cropping, such as leakage of nutrients to ground and surface waters along with atmospheric emissions. Nitrogen fertilizer recommendations for warm-season perennial grasses, such as switchgrass suggest a range of 10 to 12 kg ha^{-1} of N is required for each Mg of biomass harvested [50]. Crop rotations that include legumes to fix atmospheric N or mixtures of grasses and legumes potentially could reduce fertilizer N inputs into bioenergy cropping systems and improve their energy balance.

Warm-season perennial grasses internally recycle N from the aboveground shoots to the crown and roots in the fall for use in overwintering and regrowth the following spring [51]. This mechanism enables an efficient use and reuse of N by the plant. *Miscanthus* had lower N fertilizer demand than reed canarygrass because of an efficient internal N recycling mechanism [30]. We lack critical information on when N recycling occurs within the plant, how much N recycles among plant organs, and quantitative data on how much recycling contributes to the N economy of a biomass energy crop.

Harvest management for biomass feedstock emphasizes yield and persistence. Producers managing stands that are dedicated to production of biomass feedstock may want harvest flexibility to respond to potential fluctuations in future feedstock markets [52]. Time of harvest affects switchgrass yield and varies with agroecological region. In the south-central USA, a single harvest in mid-September maximized biomass yields [21]. In the Central Plains of the USA, harvesting switchgrass at maturity (mid- to late August) maximized biomass yield. Delaying harvest until after a killing frost in October, however, reduced yields by 10 to 20% [53]. In the northcentral USA a mid-August harvest reduced stand density compared with a fall harvest [54]. In the northeastern USA, mid- to late-summer annual yields were similar to spring-harvest yields after the initial high yield the first year of summer harvest [55]. Time of harvest also affects feedstock quality. In direct combustion systems, minerals in biomass can corrode and foul boilers [56]. The ash concentration of switchgrass decreases as it matures [57] leading to improved utility for conversion and potentially lower N requirements with a fall vs. summer harvest [53]. Delaying biomass harvest from fall to spring reduces mineral and water concentration of perennial forage grasses, but may also reduce yields [55].

SET-ASIDE, MARGINAL, AND ABANDONED GRASSLANDS AS BIOMASS FEEDSTOCK RESOURCES

In addition to using new plantings of perennial bioenergy crops, other grassland-based resources such as set-aside lands along with marginal or abandoned lands could supply bioenergy feedstock. For example, cool-season grassland pasture on marginal lands in southern Iowa USA produced 0.8 to 8.2 Mg biomass ha^{-1} depending on site characteristics and environment (Table 1) [58].

Land in the Conservation Reserve Program ((CRP); a government land set-aside program) has been suggested as a potential resource for biomass feedstock in the USA [59, 60]. The goal of the CRP is to remove land from crop production and establish long-term vegetation cover to prevent soil erosion, improve water quality, and enhance wildlife habitat. Of the 13.8 million ha of CRP land in the USA, about 6.8 million ha are potentially available for biomass feedstock production [59]. In the Northeastern USA, a two-year study demonstrated a standing feedstock supply of 6.6 Mg ha^{-1} on CRP and other conservation lands (Table 1) [61].

Maintaining the environmental benefits of the CRP is a concern when considering its potential for bioenergy production. This would include maintaining a perennial vegetative cover to prevent soil erosion and judiciously using fertilizers to obtain economic yields and not compromise water quality.

Management practices suggested by research on CRP grasslands in the northern Great Plains of the USA included harvesting after a killing frost and adding fertilizer N at 56 kg ha^{-1} to maintain biomass production and stand persistence [62]. Biomass yields ranged from 2 to 8 Mg ha^{-1} at two locations during 3 yr. Carbon sequestration to a 90-cm depth on these same lands ranged from 2.4 Mg C ha^{-1} yr^{-1} with inorganic N fertilizer to 4.0 Mg C ha^{-1} yr^{-1} with manure application [63]. Mapemba et al. [64] applied economic models to estimate the cost of delivering a flow of feedstock from CRP lands in the southern Great Plains of the USA. Estimated feedstock costs ranged from US$29 to US$64 Mg^{-1} depending on conversion plant size, frequency of harvests, and the number of days suitable for harvest.

Other management considerations for the use of CRP lands for biofuels in the future would include harvest management consistent with maintaining wildlife habitat.

Low-Input High-Diversity Prairie Systems

An intriguing development in using perennial crops for biomass is the so-called low-input highdiversity (LIHD) prairie approach [65]. Based on previous research into the productivity-biodiversityecosystem function relationship of ecosystems, estimates were made that low-input prairies could provide more usable energy and greater environmental benefits than corn-grain (*Zea mays* L.) ethanol or soybean (*Glycine max* L. Merr) biodiesel. Total CO_2 sequestration (soil plus roots) rate under the high-diversity plots in Cedar Creek, Minnesota was 4.4 Mg CO_2 ha^{-1} yr^{-1}.

The LIHD approach relies on extremely low or no inputs of fertilizers and other materials and instead exploits increased productivity of multispecies plant communities on degraded lands. Criticisms of the research pointed out that the research was done under highly artificial conditions, (very small plots, hand weeding, and extremely high labor inputs) and did not represent realistic conditions [66]. These criticisms were rebutted [67]. Others, however, question whether degraded lands have the ability to produce the abundant quantities of biomass needed [68]. On-farm research on the northern Great Plains of the USA in large fields (3 to 9 ha) demonstrated that management of switchgrass as a bioenergy crop along with modest inputs of N fertilizer produced 93% more net energy than LIHD plots in Minnesota [49]. Switchgrass biomass yields in on-farm fields ranged from 5 to 11 Mg ha^{-1}, whereas LIHD plots averaged 3.6 Mg ha^{-1} (Table 1). Soil C sequestration under switchgrass managed for bioenergy on the Northern Plains averaged 4.42 Mg C ha^{-1} yr^{-1} [69].

Switchgrass in monoculture performed poorly on the degraded soils of the Cedar Creek site; however, switchgrass in monoculture and binary mixture with a legume has been highly productive (5 to 16 Mg biomass ha^{-1} over 13 yr) on extremely degraded soils of abandoned strip mines in Virginia [70].

Integrated Bioenergy Crop-Production Systems

A perennial crop permanently dedicated to biomass feedstock production would seem to be an ideal goal because (1) there would be no annual re-establishment costs, (2) tillage would be eliminated, which would reduce inputs, costs, and soil erosion, and (3) a permanent vegetative cover would sustain soil conservation and water-quality protection. Perennials, however, are rarely permanent and some annual cropping or innovative combinations of annual and perennial bioenergy crops strategically deployed across the farm landscape and combined into synergistic rotations may be necessary in the future [71].

Combining annual bioenergy crops such as corn and sorghum (*Sorghum bicolor* L. Moench.) into rotations with perennial bioenergy crops, perhaps to jump-start the establishment phase, may benefit bioenergy cropping systems [71]. Including new cover crops, such as kura clover (*Trifolium ambiguum* L.), with annual crop systems as "living mulches" [72] could reduce environmental impacts.

Innovative combinations of cool-season and warm-season annual crops could be the basis for dedicated biomass double cropping. Double cropping cool-season cereals with warm-season annuals, such as corn and sorghum was successful in the Midwestern USA if the cereal crop was removed as forage early in the spring to allow early planting of corn and sorghum for biomass [73].

Results from cropping-systems research in southern Germany indicated that perennial biomass systems based on *Miscanthus*, switchgrass, or willows (Salix schernii E. Wolf x viminalis) could be as productive as energy maize with lower energy inputs (Figure 1) [34]. Nitrogen fertilizer was the most energy-intensive input and accounted for 41 to 64% of energy inputs for annual crops and 17 to 45% of inputs for perennials. Energy maize (grown with a grass cover crop for erosion and nutrient control) had the greatest land use efficiency because of high yields (19.1 Mg ha^{-1} of grain+stover), whereas the perennial crops had the greatest N use efficiency (Figure 1). Crop rotations based on annual coolseason plants [wheat (*Triticum aestivum* L.), triticale (Triticosecale x Wittmack), and oilseed rape (*Brassica napus* L.)] had the lowest energy use efficiency. Reducing cultivation intensity via no-tillage management in the cool-season crop rotations had a small effect on cropping system energy savings.

Relying on a diversity of crops and cropping systems in farm landscapes and larger scales (watersheds) would endow future bioenergy production systems with greater stability, resistance, and resilience to climatic and other environmental shocks [74, 75].

ECONOMICS AND ENVIRONMENT

Currently, bioenergy produced from second generation cellulosic feedstocks costs more than fossil fuels [1]. Biomass yield, harvest and transport costs, conversion efficiency, and cost of fossil fuel used to produce the feedstock determine the economics of bioenergy production and vary among regions of the USA. Projected costs of switchgrass biomass range from \$40 to \$61 Mg^{-1} in Tennessee and \$53 to \$74 Mg^{-1} in Oklahoma [76]. Target costs estimated for economical production of ethanol from biomass feedstocks are about \$38 Mg^{-1} [60]. The value of environmental benefits of bioenergy crops, however, may offset the price differential between biofuels and fossil fuels [10]. Results from watershed modeling studies suggest that payments of \$11 to \$27 Mg^{-1} of switchgrass biomass could be justified on the basis of projected water quality benefits from reduced sedimentation, erosion, and nutrient losses associated with growing a perennial grass [77].

The environmental benefits of bioenergy crops include increased soil quality, reduced losses of soil nutrients, soil C sequestration, protecting riparian zones [78], and recycling nutrients from sewage sludge, livestock manure, and bioconversion byproducts [71, 79-81] among others.

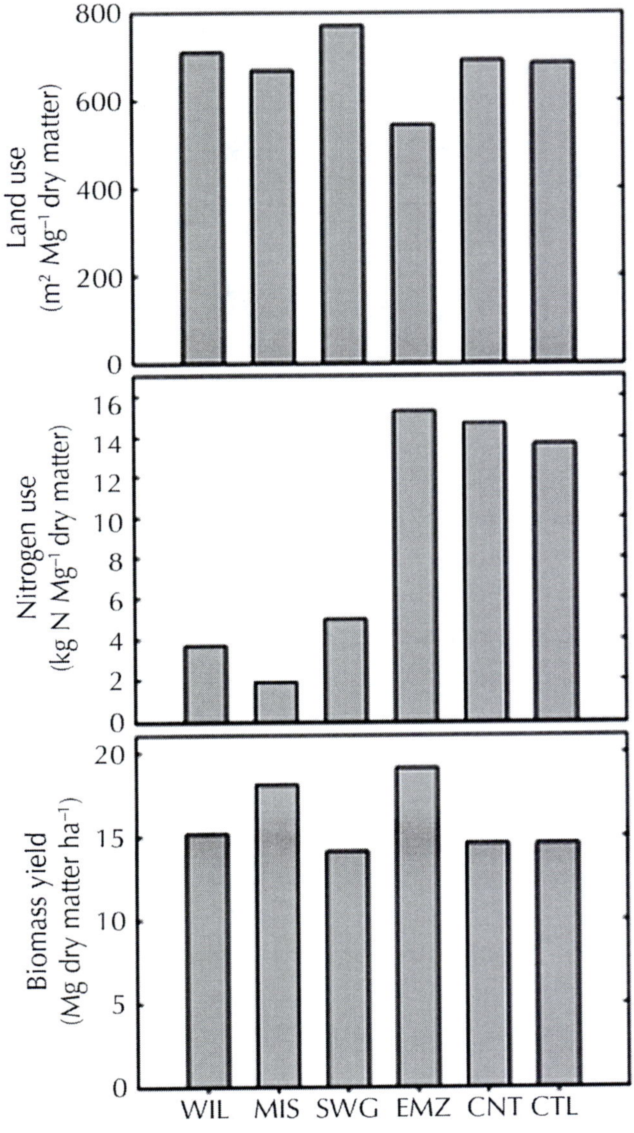

Figure 1: Biomass yield, nitrogen use efficiency, and land use efficiency for six cropping systems compared during 4 yr in southern Germany [34]. WIL, willow trees+27 kg N ha^{-1}; MIS, *Miscanthus*+40 kg N ha^{-1}; SWG, switchgrass+80 kg N ha^{-1}; EMZ, energy maize +120 kg N ha^{-1} and grown with a grass cover crop; CNT, winter wheat, winter triticale, and oilseed rape grown in rotation with no-till methods; CTL, same crop rotation as CNT but grown with conventional tillage methods.

In the Chesapeake Bay region of the USA, concern over the potential for large-scale conversion of land to corn production for biofuels, with the potential attendant increase in soil erosion and transport of nutrients to the bay prompted the Chesapeake Bay Commission to explore alternative scenarios for bioenergy cropping in the Chesapeake Bay watershed [82]:

1. 121, 500 additional ha of corn production under typical management
2. 121, 500 additional ha of soybean production under typical management
3. 121, 500 ha of switchgrass (converted from pasture and hayland) planted for perennial biomass energy crop with no nitrogen fertilizer
4. 121,500 ha of additional corn added but produced with best management practices such as cover crops and other technologies that reduce erosion and nutrient losses.
5. 405,000 ha of switchgrass (converted from pasture and hayland) planted for perennial biomass energy crop with no nitrogen fertilizer

The estimated changes in N load to the bay for each of the scenarios showed that simply adding more row crops with typical management increased N loading to the bay (Figure 2). However, replacing pasture and hayland with switchgrass grown without nitrogen fertilizer substantially reduced N loading as did adopting new BMPs and cropping practices for corn. Switchgrass, however, requires some nitrogen fertilizer input for production [50]. Therefore, growing switchgrass for biomass production with no nitrogen fertilizer input would be impractical.

The increased use of biomass energy crops has been recommended as a strategy for mitigating atmospheric increases in CO_2 [83]. Advocates of biofuels consider them to have a near-zero net emission of GHGs because they recycle carbon, consuming it during crop production and releasing it as the fuels are used, as compared with fossil fuels, which release ancient carbon that was consumed by plants long ago. However, growing the crops requires energy (e.g., to operate farm machinery), as does converting the harvest into usable fuels. In the near term, CO_2 can be sequestered in the soil during plant growth, thereby reducing GHG concentrations; in the long term, however, the soil's capacity to store carbon is limited. Along with soil carbon sequestration, coproducts of biofuel production, such as lignin and protein, can displace net GHG emissions, making these system sinks "carbon- or GHG-negative." Coproducts can "remove CO_2" from the atmosphere by displacing demand for fossil fuels, so as new technology is developed to extract more coproducts from biomass, this sink for CO_2 will increase.

Increases in soil organic C under warm-season perennial grasses have been reported in several regions of the USA [84-87] and central Europe [88]. Potential benefits from C sequestration under perennial grasses depend on the cropping system used or replaced. In the southeastern USA, switchgrass grown for biomass accumulated more total soil C than cotton (Gossypium Hirsutum L.) or corn row crops, but not as much as bahiagrass (Paspalum notatum Flugge) pasture [89]. Studies in Tennessee and Virginia, however, have shown that soil organic C concentrations and inventory under switchgrass were no different than beneath tall fescue (Festuca Arundinacea Schreb.) or pasture [90].

Short-term (5- to 7-yr) soil C gains under switchgrass were minimal when it was managed for hay or grazing in Pennsylvania [91].

Greenhouse Gas Emissions from Perennial Bioenergy Crops

Agriculture is responsible for about 7% of total USA GHG emissions [92]. Proper management of agricultural systems can reduce direct emissions and offset emissions from other entities by sequestering C in the soil [93-95]. However, GHG emissions from USA forage and grazing lands have not been extensively quantified [96].

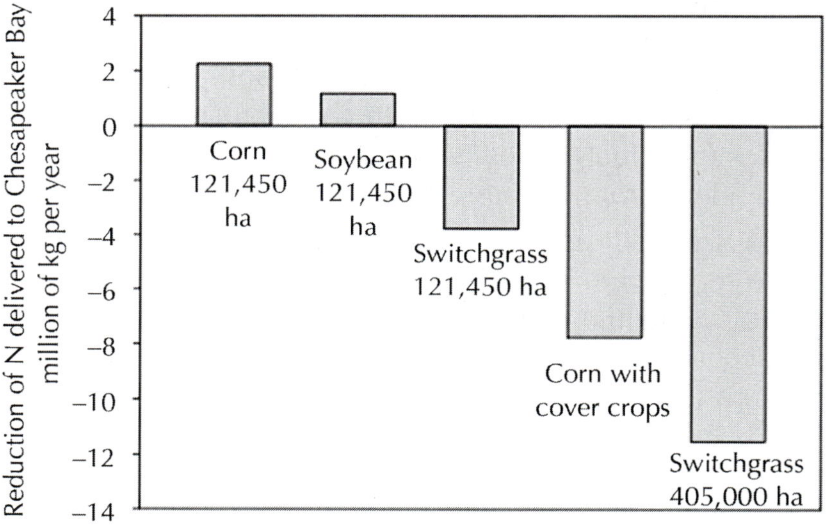

Figure 2: Estimated change in nitrogen loading to the Chesapeake Bay resulting from five cropping system scenarios: (1) 121, 500 additional ha of corn production under typical management; (2) 121, 500 additional ha of soybean production under typical management; (3) 121, 500 ha of switchgrass (converted from pasture and hayland) planted for perennial biomass energy crop with no nitrogen fertilizer; (4) 121,500 ha of additional corn added added but produced with best management practices such as cover crops and other technologies that reduce erosion and nutrient losses; (5) 405,000 ha of switchgrass (converted from pasture and hayland) planted for perennial biomass energy crop with no nitrogen fertilizer. Adapted from the Chesapeake Bay Commission [82].

Biofuels have a large potential to reduce the GHG emissions associated with energy use [5].

Bioenergy cropping systems could help offset GHG emissions, but quantifying that offset is complex. Bioenergy crops offset CO_2 emissions by converting atmospheric CO_2 to organic C in crop biomass and soil, but they also emit N_2O and vary in their effects on soil oxidation of methane. Quantifying these factors is necessary to determine the net effect of several bioenergy cropping systems on soil C sequestration and GHG emissions.

In bioenergy cropping systems, N_2O emissions were the largest source of GHGs [5]. Nitrous oxide emissions result from N inputs from fixation, fertilization, above-ground residue, decomposition of below-ground residue, and mineralization of soil organic matter [97]. Management practices, such as tillage, also affect N_2O emissions from the soil [98]. Fossil fuel inputs from agricultural machinery and chemical application are affected by both the choice of crop and management practices. Reducing farm operations through reducing tillage, planting, and N fertilizer use significantly reduced net GHG emissions [5, 94, 99, 100].

Accounting for the effects of global land use change in GHG emission calculations, however, significantly alters estimates of the global warming potential of bioenergy crops. Converting land currently in forest or grassland to bioenergy crop production incurs a large "carbon debt" that may take decades or centuries to repay [101]. Model simulations indicate that producing grassland biomass on marginal or degraded lands does not incur a carbon debt [101].

Other computer simulation analyses of land use change suggest that any diversion of cropland from food to bioenergy production in the USA will result in more land converted to food crop production in other countries. The associated land clearing and crop production will actually increase GHG emissions [68]. Thus, policy analysts recommend that emphasis be placed on using crop residues, municipal solid wastes, and other biomass sources that use carbon otherwise not incorporated into human food.

Energy Balance Metrics of Bioenergy Cropping

Life cycle assessment has been used to evaluate the environmental impacts of products through quantifying their energy and material flows at all stages [102]. The energy balance and a consideration of N cycles are some of the key elements in life cycle assessments of biofuels. The energy balance for switchgrass production considers the energy content of the biomass minus the fossil energy used in production (i.e., the net energy production from the system). Biomass can be directly combusted or the cellulose fraction can be converted to ethanol and the lignin fraction combusted. Producing ethanol from switchgrass results in an energy ratio (ratio of energy output vs. energy input; values greater than 1 imply energy output greater than input) of about 5.4 compared with 1.25 from corn grain [49]. This is because the corn stover is not included in the energy balance. Return of corn stover to the soil is needed to reduce soil erosion and maintain soil quality [103].

The N cycle has a significant impact on the energy balance and production of greenhouse gases.

The fossil fuel energy required to produce N used in biofuel production can account for a significant portion of the total system energy requirements. Thus, reducing the amount of fertilizer N needed can significantly increase the energy balance (net production) of the system. Perennial forage crops can have lower requirements for N than an annual crop like corn, thereby reducing the fossil fuel energy requirement. Lower N use also reduces the emission of N_2O, a potent greenhouse gas.

IMPLICATIONS OF LARGE SCALE BIOENERGY CROPPING WITH PERENNIAL FORAGES

In the USA, national goals for renewable energy call for 20% of transportation fuels, 25% of chemicals, and 5% of the nation's power to come from biomass feedstock by 2030 [104]. Achieving these targets will require an annual biomass supply of 907 million Mg (1 billion tons) [60]. The biomass would come from crop residues, perennial energy crops, manures and other waste materials, and grains. The DOE-USDA group estimated that 22.3 million ha of the 181.4 million ha of USA cropland would be needed to produce perennial biomass feedstock. The land area was proposed to come from cropland pasture (9.1 million ha; permanent pastureland was excluded from the analysis), hayland (4.2 million ha; but not alfalfa hayland), CRP (4 million ha; only those acres suited to grass production), and reallocation of existing cropland (5.7 million ha).

There were several assumptions made to meet those estimates. Key among them were (1) grain yields would increase by 50% by the year 2030; (2) technology will be developed to recover 75% of all annual crop residues; (3) all cropland would be managed via no-till; and (4) average annual biomass yields of 12.3 Mg ha^{-1} across all perennial bioenergy crop lands were achievable. Figure 3 illustrates the yield increases needed to reach assumption 4.

The billion-ton report also contained some very important caveats acknowledged by the authors, including the following (1) demand for meat production could increase, which would make conversion of cropland to perennial energy crops less likely; (2) higher export demand for some crops could limit cropland conversion to perennial crops; (3) the expected demand for forage would likely decrease because of the current trend of increasing the proportion of cattle in large confined animal feeding operations.

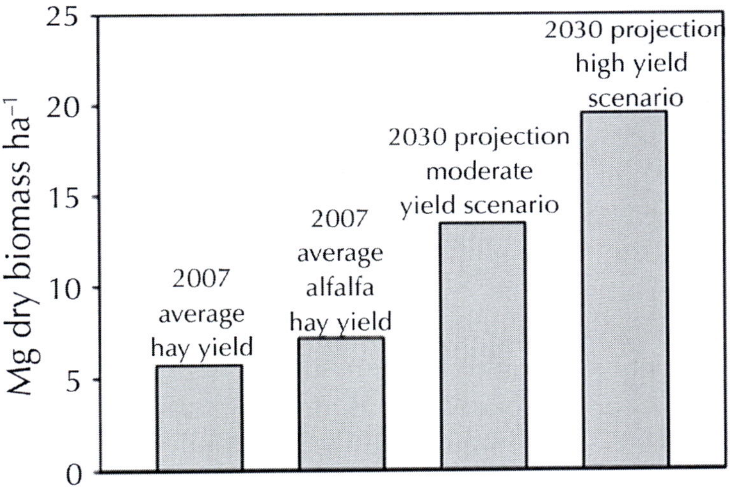

Figure 3: Comparison of 2007 average hay yields (Mg ha^{-1}) in the USA with herbaceous biomass yields projected for perennial energy crops in 2030. Current hay yields are from the National Agricultural Statistics Service (www.nass.usda.gov) and projected 2030 biomass yields are from the "billion ton" report of Perlack et al. [60].

The assumptions and caveats have important implications for how existing forage and grazing lands are used in the future. For example, if the targets and technology are not met for grain yields and crop residue removal then much more land area will be required to produce the billion tons of biomass. Similarly, if the first two caveats reduce the likelihood of cropland conversion to perennial energy crops, then the production of perennial energy crops could be forced to more marginal lands. Similarly, the proposed replacement of 405,000 ha of pasture and hayland with switchgrass in the Chesapeake Bay region of the USA [82] would force forage-livestock production to other regions or cause greater intensification of confined animal production. Projections of expanded ethanol and biodiesel production in the USA to 60 billion gallons in 2030 indicate large reductions in pastureland acreage [105]. All of these aspects will place tremendous pressures on hay, forage, and pastureland in the future and the expanding land base necessary for biomass production would probably force forage and grazing lands production to ever more marginal lands. This could have very important implications for the forage-livestock industry. Shortages of forage crops and land for grazing livestock could result in higher production costs and reduce the profitability of livestock production. For example, De la torre Ugarte et al. [59] outlined such a scenario: with the introduction of bioenergy crops, traditional crops may lose acreage to the new crops, which would generate higher prices for the traditional crops and in turn there would be incentives to convert more idle land and pastureland to traditional crop production. Progressing to the second generation of biofuels will require transitions in the forage-livestock industry and in agriculture as a whole to accommodate both fuel and food production. Perennial forages used as biomass feedstock crops are a key component of this transitional process. Farmers face several competing demands and pressures from markets, governmental policies, and society expectations in producing crops for food or fuel. It is unlikely that the quest for renewable fuels will end with the second generation. Developing and implementing renewable energy production systems based on sound environmental stewardship should be the legacy passed on by the second generation to future generations of renewable fuels.

REFERENCES

1. Lange, J-P. 2007. Lignocellulose conversion: an introduction to chemistry, process, and economics. Biofuels, Bioproducts, and Biorefineries 2007, 1, 39-48.

2. Samson, R.; Mani, S.; Boddey, R.; Sokhansanj, S.; Quesada, D.; Urquiaga, S.; Reis, V.; Ho Lem, C. The potential of C_4 perennial grasses for developing a global BIOHEAT industry. Critical Reviews in Plant Science 2005, 24, 461-495.

3. McLaughlin, S.B.; Kiniry, J.R.; Taliaferro, C.M.; De La Torre Ugarte, D. Projecting yield and utilization potential of switchgrass as an energy crop. Advances in Agronomy 2006, 90, 267-297.
4. Sanderson, M.A.; Adler, P.R.; Boateng, A.A.; Casler, M.D.; Sarath, G. Switchgrass as a biofuels feedstock in the USA. Canadian Journal of Plant Science 2006b, 86, 1315-1325.
5. Adler, P.R.; Del Grosso, S.J.; Parton, W.J. Life cycle assessment of net greenhouse gas flux for bioenergy cropping systems. Ecological Applications 2007, 17, 675–691.
6. Parrish, D.; Fike, J.H. The biology and agronomy of switchgrass for biofuels. Critical Reviews in Plant Science 2005, 24, 423-459.
7. Sanderson, M.A.; Adler, P.R.; Martin, N.P. Biomass, energy, and industrial uses of forages. In Forages Volume II: The science of grassland agriculture, 6th Ed.; Barnes, R.F.; Nelson, C.J., Moore, K.J.; Collins, M., Eds.; Iowa State University Press: Ames, IA, 2007; pp. 635-647.
8. Vogel, K.P. Energy production from forages (or American agriculture--back to the future). Journal of Soil and Water Conservation 1996, 51, 137-139.
9. Barnes, R.F; Nelson, C.J. Forage and grasslands in a changing world. In Forages Volume I: An introduction to grassland agriculture, 6th Ed.; Barnes, R.F.; Nelson, C.J., Moore, K.J., Eds. Iowa State University Press: Ames, IA, 2003; pp. 3-23.
10. McLaughlin, S.B.; de la Torre Ugarte, D.G.; Garten, jr, C.T.; Lynd, L.R.; Sanderson, M.A.; Tolbert, V.G.; Wolf, D.D. High-value renewable energy from prairie grasses. Environmental Science and Technology 2002, 36, 2122-2129.
11. Szego, G.C.; Kemp, C.C. Energy forests and fuel plantations. Chemical Technology 1976, 3, 275-284.
12. Povich, M.J. Fuel farming. Chemical Technology 1976, 6, 434-439.
13. Helsel, Z.R.; Wedin, W.F. Direct combustion energy from crops and crop residues produced in Iowa. Energy in Agriculture 1981, 1, 317-329.
14. Wedin, W.F.; Helsel, Z.R. Plant species for biomass production on marginal sites. In Proc. Conf. on Production from grasses, crop

residues, forage legumes, and grasses for conversion to fuels and chemicals; Midwest Research Institute, Kansas City, MO, March 2-3, 1977; p. 34-48

15. Ragauskas, A.J.; Williams, C.K.; Davidson, B.H.; Britovsek, G.; Cairney, J.; Eckert, C.A.; Frederick, Jr., W.J; Hallett, J.P; Leak, D.J.; Liotta, C.L.; Mielenz, J.R.; Murphy, R.; Templer, R.; Tschaplinski, T. The path forward for biofuels and biomaterials. Science 2006, 311, 484-489.
16. Boody, G.; Vondracek, B.; Andow, D.A.; KinKrinke, M.; Westra, J.; Zimmerman, J.; Welle, P. Multifunctional agriculture in the United States. Bioscience 2005, 55, 27-38.
17. Keeney, D.R.; Sanderson, M.A. Forages and the environment. In Forages Volume II: The science of grassland agriculture, 6th Ed.; Barnes, R.F.; Nelson, C.J., Moore, K.J.; Collins, M., Eds., Iowa State University Press: Ames, IA, 2007; pp. 167-176.
18. Barnes, R.F; Nelson, C.J.; Moore, K.J.; Collins, M. Forages Volume II: The science of grassland agriculture, 6th Ed.; Iowa State University Press, Ames, IA, 2007.
19. McLaughlin, S.B.; Kzos, L.A. Development of switchgrass (Panicum virgatum) as a bioenergy feedstock in the United States. Biomass and Bioenergy 2005, 28, 515-535.
20. Bouton, J.H. Molecular breeding of switchgrass for use as a biofuel crop. Current Opinion in Genetics & Development 2007, 17, 553-558.
21. Sanderson, M.A.; Reed, R.L; Ocumpaugh, W.R; Hussey, M.A; Van Esbroeck, G.; Read, J.C.; Tischler, C.R; Hons, F.M. Switchgrass cultivars and germplasm for biomass feedstock production in Texas. Bioresource Technology 1999, 67, 209-219.
22. Vogel, K.P.; Hopkins, A.A.; Moore, K.J; Johnson, K.D.; Carlson, I.T. Registration of 'Shawnee' switchgrass. Crop Science 1996, 36, 1713.
23. Burns, J.C.; Godshalk, E.B.; Timothy, D.H. Registration of 'Performer' switchgrass. Journal of Plant Registrations 2008, 2, 29-30.
24. Burns, J.C.; Godshalk, E.B.; Timothy, D.H. Registration of 'BoMaster' switchgrass. Journal of Plant Registrations 2008, 2, 31-32.

25. Casler, M.D.; Stendal, C.A.; Kapich, L.; Vogel, K.P. Genetic diversity, plant adaptation regions, and gene pools for switchgrass. Crop Science 2007, 47, 2261-2273.

26. Vogel, K.P.; Jung, H.G. Genetic modification of herbaceous plants for feed and fuel. Critical Reviews in Plant Science 2001, 20, 15-49.

27. Rose, L.W.; Das, M.K; Taliaferro, C.M. Estimation of genetic variability and heritability for biofuel feedstock yield in several populations of switchgrass. Annals of Applied Biology 2008, 152, 11-17.

28. Lewandowski, I.; Scurlock, J.M.O.; Lindvall, E.; Christou, M. The development and current status of perennial rhizomatous grasses as energy crops in the US and Europe. Biomass and Bioenergy 2003, 25, 335-361.

29. Chapman, G.P. The biology of grasses; CAB International Press: Wallingford, UK, 1996.

30. Lewandowski, I.; Schmidt, U. Nitrogen, energy, and land use efficiencies of *miscanthus*, reed canary grass and triticale as determined by the boundary line approach. Agriculture, Ecosystems, and Environment 2006, 112, 335-346.

31. El-Bassam, N. Energy plant species: their use and impact on environment and development. James and James Science Publishers: London, UK, 1998.

32. Heaton, E.; Voigt, T; Long, S.P. A quantitative review comparing the yields of two candidate C_4 perennial biomass crops in relation to nitrogen, temperature, and water. Biomass and Bioenergy 2004, 27, 21-30.

33. Heaton, E.A.; Clifton-Brown, J.; Voigt, T.B.; Jones, M.B.; Long, S.P. *Miscanthus* for renewable energy generation: European Union experience and projections for Illinois. Mitigation and Adaptation Strategies for Global Change 2004, 9, 433-451.

34. Boehmel, C.; Lewandowski, I; Claupein, W. Comparing annual and perennial energy cropping systems with different management intensities. Agricultural Systems 2008, 96, 224-236.

35. Carlson, I.T.; Oram, R.N.; Suprenant, J. Reed canarygrass and other Phalaris species. In Coolseason forage grasses; Moser, L.E.;

Buxton, D.R.; Casler, M.D, Eds.; American Society of Agronomy, Madison, WI, 1996; Agronomy Monograph 34, pp. 569-604

36. Cherney, J.H.; Johnson, K.D.; Lechtenberg, V.L.; Hertel, J.M. Biomass yield, fiber composition, and persistence of cool-season perennial grasses. Biomass 1986, 10, 175-186.
37. Hallam, A.; Anderson, I.C.; Buxton, D.R. Comparative economic analysis of perennial, annual, and intercrops for biomass production. Biomass and Bioenergy 2001, 21, 407-424.
38. Partala, A.; Mela, T.; Esala, M.; Ketoja, E. Plant recovery of 15N-labelled nitrogen applied to reed canarygrass grown for biomass. Nutrient Cycling in Agroecosystems 2001, 61, 273-281.
39. Merigliano, M.F.; Lesica, P. The native status of reed canarygrass (Phalaris arundinacea L.) in the inland northwest, USA. Natural Areas Journal 1998, 18, 223-230.
40. Russelle, M.P. Alfalfa. American Scientist 2001, 89, 252-261.
41. Delong, M.M.; Swanberg, D.R.; Oelke, E.A.; Hanson, C.; Onishak, M.; Schmid, M.R.; Wiant. B.C. Sustainable biomass energy production and rural economic development using alfalfa as a feedstock. In Proceedings of the 2nd Biomass Energy Conference of the Americas; D.L. Klass, Ed.; Portland, OR, August 21-24, 1995; pp. 1582-1591
42. Lamb, J.F.S.; Sheaffer, C.C.; Samac, D.A. Population density and harvest maturity effects on leaf and stem yield in alfalfa. Agronomy Journal 2003, 95, 635-641.
43. Lamb, J.F.S.; Jung, H.G.; Sheaffer, C.C.; Samac, D.A. Alfalfa leaf protein and stem cell wall polysaccharide yields under hay and biomass management systems. Crop Science 2007, 47, 1407-1415.
44. Prine, G.M.; Stricker, J.A.; Anderson, D.L.; Windham, W.R. Chemical composition of biomass from tall perennial tropical grasses. In Proceedings of the 2nd Biomass Energy Conference of the Americas; D.L. Klass, Ed.; Portland, OR, August 21-24, 1995; pp. 278-283.
45. Boateng, A.A.; Anderson, W.F.; Phillips, J.G. Bermudagrass for biofuels: effect of two genotypes on pyrolysis product yield. Energy and Fuels 2007, 21, 1183-1187.
46. Schank, S.C.; Chenowyth, D.P; Turick, C.E.; Mendoza, P.E.

Napiergrass genotypes and plant parts for biomass energy. Biomass and Bioenergy 1993, 4, 1-7.

47. Springer, T.E.; Dewald, C.L. Eastern gamagrass and other Tripsacum species. In Warm-season (C_4) grasses; Moser, L.E.; Burson, B.L.; Sollenberger, L.E., Eds.; American Society of Agronomy, Madison, WI, 2004; pp. 955-973.

48. Boe, A.; Lee, D.K. Genetic variation for biomass production in prairie cordgrass and switchgrass. Crop Science 2007, 47, 929-934.

49. Schmer, M.R.; Vogel, K.P; Mitchell, R.B.; Perrin, R.K. Net energy of cellulosic ethanol from switchgrass. Proceeding National Academy of Science 2008, 105, 464-469.

50. Brejda, J.J. Fertilization of warm-season grasses. In Native warm-season grasses: research trends and issues; Moore, K.J., Anderson, B.E., Eds.; Crop Science Society of American: Madison, WI, 2000; Special Publication No. 30, pp. 177-200

51. Clark, F.E. Internal cycling of 15N in shortgrass prairie. Ecology 1977, 58, 1322-1333.

52. Anderson, W.F.; Dien, B.S.; Brandon, S.K.; Peterson, J.D. Assessment of bermudagrass and bunch grasses as feedstock for conversion to ethanol. Applied Biochemistry and Biotechnology 2007, DOI: 10.1007/s12010-007-8041-y.

53. Vogel, K.P.; Brejda, J.J.; Walters, D.T.; Buxton, D.R. Switchgrass biomass production in the midwest: Harvest and nitrogen management. Agronomy Journal 2002, 94, 413-420.

54. Casler, M.D.; Boe, A.R. Cultivar × environment interactions in switchgrass. Crop Science 2003, 43, 2226-2233.

55. Adler, P.R.; Sanderson, M.A.; Boateng, A.; Weimer, P.J.; Jung, H.G. Biomass yield and biofuel quality of switchgrass harvested in fall or spring. Agronomy Journal 2006, 98, 1518-1525.

56. Jenkins, B.M.; Baxter, L.L; Miles Jr., T.R.;. Miles, T.R. Combustion properties of biomass. Fuel Processing Technology 1998, 54, 17-46.

57. Sanderson, M.A.; Wolf, D.D. Switchgrass biomass composition during morphological development in diverse environments. Crop Science 1995, 35, 1432-1438.

58. Florine, S.E.; Moore, K.J.; Fales, S.L.; White, T.A.; Burras, C.L. Yield and composition of herbaceous biomass harvested from naturalized grassland in southern Iowa. Biomass and Bioenergy 2006, 30, 522-528.
59. De La Torre Ugarte, D.G.; Walsh, M.; Shapouri,H.; Slinsky, S.P. The economic impacts of bioenergy crop production on U.S. agriculture. USDA Office of Energy Policy and New Uses. Agricultural Economics Report No. 816. Washington, D.C., 2003.
60. Perlack, R.D.; Wright, L.L.; Turhollow, A.F.; Graham, R.L.; Stokes, B.J; Erbach, D.C. Biomass as a feedstock for a bioenergy and bioproducts industry: The technical feasibility of a billion-ton annual supply. A joint study sponsored by the U.S. Dept. Energy and the USDA, 2005; www1.eere.energy.gov/biomass/publications.html (Accessed March 15, 2008).
61. Adler, P.R.; Sanderson, M.A.; Goslee, S.C. Management and composition of conservation lands in the Northeastern United States. In Proceedings of the 4th Eastern Native Grass Symposium, Lexington, KY, October 3-6, 2004; pp. 187-198, www.uky.edu/Ag/Forestry/TBarnes/Assets/Proceeding.pdf (Accessed on March 15, 2008).
62. Mulkey, V.R.; Owens, V.N.; Lee, D.K. Management of switchgrass-dominated conservation reserve program lands for biomass production in South Dakota. Crop Science 2006, 46, 712-720.
63. Lee, D.K.; Owens, V.N.; Doolittle, J.J. Switchgrass and soil carbon sequestration response to ammonium nitrate, manure, and harvest frequency on conservation reserve program land. Agronomy Journal 2007, 99, 462-468.
64. Mapemba, L.; Epplin, F.M.; Taliaferro, C.M.; Huhnke, R.L. Biorefinery feedstock production on conservation reserve program land. Review of Agricultural Economics 2007, 29, 227-246.
65. Tilman, D.; Hill, J.; Lehman, C. Carbon-negative biofuels from low-input high-diversity grassland biomass. Science 2006, 314, 1598-1600.
66. Russelle, M.P.; Morey, R.V.; Baker, J.M.; Porter, P.M.; Jung, H.G. Comment on "carbon negative biofuels from low-input high-diversity grassland biomass." Science 2007, 316, 1576b.

67. Tilman, D.; Hill, J; Lehman, C. Response to "Comment on carbon negative biofuels from lowinput high-diversity grassland biomass." Science 2007, 316, 1576c.
68. Searchinger, T.; Heimlich, R; Houghton, R.A.; Dong, F.; Elobeid, A.; Fabiosa, J; Tokgoz, S.; Hayes, D.; Yu, T. Use of U.S. croplands for biofuels increases greenhouse gases through emissions from land use change. Science 2008, 319, 1238-1240.
69. Liebig, M.A.; Johnson, H.A.; Hanson, J.D.; Frank, A.B. Soil carbon under switchgrass stands and cultivated crop land. Biomass and Bioenergy 2005, 28, 347-354.
70. Evanylo, G.K.; Abaye, A.O.; Dundas, C.; Zipper, C.E.; Lemus, R.; Sukkariyah, B.; Rockett, J. Herbaceous vegetation production, persistence, and metals uptake on a biosolids-amended mine soil. Journal of Environmental Quality 2005, 34, 1811-1819.
71. Anex, R.P.; Lynd, L.R.; Laser, M.S.; Heggenstaller, A.H.; Liebman, M. Potential for enhanced nutrient cycling through coupling of agricultural and bioenergy systems. Crop Science 2007, 47, 1327-1335.
72. Zemenchek, R.A.; Albrecht, K.A.; Boerboom, C.M; Lauer, J.G. Corn production with kura clover as a living mulch. Agronomy Journal 2000, 92, 698-705.
73. Helsel, Z.R.; Wedin, W.F. Harvested dry matter from single and double cropping systems. Agronomy Journal 1981, 73, 895-900.
74. Jordan, N.; Boody, G.; Broussard, W.; Glover, J.D.; Keeney, D.; McCown, B.H.; McIsaac, G.; Mueller, M.; Murray, M.H.; Neal, J.; Pansing, C.; Turner, R.E; Warner, K.; Wyse, D. Sustainable development of the bio-economy. Science 2007, 316, 1570-1571.
75. Kirschenmann, F.L. Potential for a new generation of biodiversity in agroecosystems of the future. Agronomy Journal 2007, 99, 373-376.
76. Epplin, F.M.; Clark, C.D.; Roberts, R.K; Hwang, S. Challenges to the development of a dedicated energy crop. American Journal of Agricultural Economics 2008, 89, 1296-1302.
77. Nelson, R.G.; Ascough II, J.C; Langemeier, M.R. Environmental and economic analysis of switchgrass production for water quality

improvement in northeast Kansas. Journal of Environmental Management 2006, 79, 336-347.

78. Tufekcioglu, A.; Raich, J.W.; Isenhart, T.M.; Shultz, R.C. Biomass, carbon and nitrogen dynamics of multi-species riparian buffers within an agricultural watershed in Iowa, USA. Agroforestry Systems 2003, 57, 187-198.

79. Sanderson, M.A.; Jones, R.M.; McFarland, M.J; Stroup, J.; Reed, R.L; Muir, J.P. Nutrient movement and removal in a switchgrass biomass-filter strip system treated with dairy manure. Journal of Environmental Quality 2001, 30, 210-216.

80. Reed, R.L.; Sanderson, M.A.; Allen, V.G; Zartman, R.E. Cadmium application and pH effects on growth and cadmium accumulation in switchgrass. Communications in Soil Science and Plant Analysis 2002, 33, 1187-1203.

81. Laird, D. The charcoal vision: A win-win-win scenario for simultaneously producing bioenergy, permanently sequestering carbon, while improving soil and water quality. Agronomy Journal 2008, 100, 178-181.

82. Chesapeake Bay Commission. Biofuels and the bay, 2007; www.chesbay.state.va.us/ Publications/BiofuelsAndTheBay1.pdf. (Accessed March 15, 2008).

83. IPCC (Intergovernmental Panel on Climate Change). IPCC Fourth assessment report, working group III, 2007; http://www.ipcc.ch/. (Accessed March 15, 2008).

84. Ma, Z.; Wood, C.W.; Bransby, D.I. Carbon dynamics subsequent to establishment of switchgrass. Biomass and Bioenergy 2000, 18, 93-104.

85. Tolbert, V.R.; Todd Jr, D.E.; Mann, L.K; Jawdy, C.M; Mays, D.R.; Malik, R.; Bandaranayake, W.; Houston, A.; Tyler, D.; Pettry, D.E. Changes in soil quality and below-ground carbon storage with conversion of traditional agricultural crop lands to bioenergy crop production. Environmental Pollution 2002, 116, S97-S106.

86. Omonode, R.A.; Vyn, T.J. Vertical distribution of soil organic carbon and nitrogen under warmseason native grasses relative to croplands in west-central Indiana, USA. Agriculture, Ecosystems, and Environment 2006, 117, 159-170.

87. Al-Kaisi, M.M.; Grote, J.B. Cropping system effects on improving soil carbon stocks of exposed subsoil. Soil Science Society of America Journal 2007, 77, 1381-1388.

88. Kahle, P.; Beuch, S.; Boelke, B.; Leinweber, P.; Schulten, H.R. Cropping of *Miscanthus* in Central Europe: biomass production and influence on nutrients and soil organic matter. European Journal of Agronomy 2001, 15, 171-184.

89. Bransby, D.I.; McLaughlin, S.B.; Parrish, D.J. A review of carbon and nitrogen balances in switchgrass grown for energy. Biomass and Bioenergy 1998, 14, 379-384.

90. Garten, C.T; Wullschleger, S.D. Soil carbon inventories under a bioenergy crop (switchgrass): Measurement limitations. Journal of Environmental Quality 1999, 28, 1359-1365.

91. Sanderson, M.A. Upland switchgrass yield, nutritive value, and soil carbon changes under grazing and clipping. Agronomy Journal 2008 100, 510-516.

92. USDA. U.S. forestry and agriculture greenhouse gas inventory: 1990-2001. Global Change Program Office, Office of the Chief Economist, U.S. Department of Agriculture. Technical Bulletin No. 1907. 163 p. March 2004; www.usda.gov/oce/global_change/gg_inventory.htm (Accessed March 15, 2008).

93. Follett, R.F.; Kimble, J.M.; Lal, R. The potential of U.S. grazing lands to sequester carbon and mitigate the greenhouse effect; CRC Press: Boca Raton, FL, 2001.

94. West, T.O.; Marland, G. A synthesis of carbon sequestration, carbon emissions, and net carbon flux in agriculture: comparing tillage practice in the United States. Agriculture, Ecosystems, and Environment 2002, 91, 261-287.

95. Lal, R.; Follett, R.F.; Kimble, J.M. Achieving soil carbon sequestration in the United States: a challenge to the policy makers. Soil Science 2003, 168, 827-845.

96. Johnson, J.M.F.; Reicosky, D.C.; Allmaras, R.R.; Sauer, T.J.; Venterea, R.T.; Dell, C.J. Greenhouse gas contributions and mitigation potential of agriculture in the central USA. Soil and Tillage Research 2005, 83, 73-94.

97. Del Grosso, S.J.; Mosier, A.R.; Parton, W.J.; Ojima, D.S. DAYCENT model analysis of past and contemporary soil N2O and net

greenhouse gas flux for major crops in the USA. Soil and Tillage Research 2005, 83, 9-24.

98. Six, J.; Ogle, S.M.; Breidt, F.J.; Conant, R.T.; Mosier, A.R.; Paustian, K. The potential to mitigate global warming with no-tillage management is only realized when practiced in the long term. Global Change Biology 2004, 10, 155-160.

99. Kim, S.; Dale, B.E. Cumulative energy and global warming impact from the production of biomass for biobased products. Journal of Industrial Ecology 2004, 7, 147-162.

100. Kim, S.; Dale, B.E. Global potential bioethanol production from wasted crops and crop residues. Biomass and Bioenergy 2004, 26, 361-375.

101. Fargione, J.; Hill, J.; Tilman, D.; Polasky, S.; Hawthorne, P. Land clearing and the biofuel carbon debt. Science 2008, 319, 1235-1236.

102. SAIC (Scientific Applications International Corporation). Life cycle assessment: principles and practice (EPA/600/R-06/060), US EPA, Office of Research and Development, National Risk Management Laboratory, Cincinnati, OH, 2006.

103. Wilhelm, W.W.; Johnson, J.M.F.; Hatfield, J.L; Voorhees, W.B.; Linden, D.R. Crop and soil productivity response to corn residue removal: A literature review. Agronomy Journal 2004, 96, 1-17.

104. Biomass Research and Development Initiative. Vision for bioenergy and biobased products in the United States, 2006; www.brdisolutions.com/publications/default.aspx (Accessed on March 15, 2008).

105. De La Torre Ugarte, D.G.; English, B.C.; and Jensen, K. Sixty billion gallons by 2030: economic and agricultural impacts of ethanol and biodiesel expansion. American Journal of Agricultural Economics 2007, 89, 1290-1295.

106. Clifton-Brown, J.C.; Lewandowski, I.; Andersson, B.; Basch, G.; Tayebi, K.; Teixeira, F.; Christian, D.G.; Riche, A.B; Bonderup-Kjeldsen, J.; Jorgensen, U.; Mortensen, J.; Schwarz, K.U. Performance of 15 *Miscanthus* genotypes at five sites in Europe. Agronomy Journal 2001, 93, 1013-1019.

107. Woodard, K.R.; Prine, G.M. Dry matter accumulation of elephantgrass, energy cane, and elephant millet in a subtropical climate. Crop Science 1993, 33, 818-824.
108. Berdahl, J.D.; Frank, A.B.; Krupinsky, J.M.; Carr, P.M.; Hanson, J.D.; Johnson, H.A. Biomass yield, phenology, and survival of diverse switchgrass cultivars and experimental strains in western North Dakota. Agronomy Journal 2005, 97, 549-555.
109. Lemus, R.; Brummer, E.C.; Moore, K.J.; Molstad, N.E.; Burras, C.L.; Barker, M.F. Biomass yield and quality of 20 switchgrass populations in southern Iowa, USA. Biomass and Bioenergy 2002, 23, 433-442.
110. Casler, M.D.; Vogel, K.P.; Taliaferro, C.M.; Elhke, N.J.; Berdahl, J.D.; Brummer, E.C.; Kallenbach, R.O.; West, C.P.; Mitchell, R.B. Latitudinal and longitudinal adaptation of switchgrass populations. Crop Science 2007, 47, 2249-2259.
111. Fike. J.H.; Parrish, D.J.; Wolf, D.D.; Balasko, J.A.; Green Jr.; J.T.; Rasnake, M.; Reynolds, J.H. Switchgrass production for the upper southeastern USA: Influence of cultivar and cutting frequency on biomass yields. Biomass and Bioenergy 2006, 30, 207-213.
112. Madakadze, I.C.; Stewart, K.; Peterson, P.R.; Coulman, B.E.; Samson, R; Smith, D.L. Switchgrass biomass and chemical composition for biofuel in Eastern Canada. Agronomy Journal 1999, 91, 696-701.
113. Alexopoulou, E.; Sharma, N.; Paptheohari, Y.; Christou, M.; Piscioneri, I.; Panoutsou, C.; Pignatelli, V. Biomass yields for upland and lowland switchgrass varieties grown in the Mediterranean region. Biomass and Bioenergy 2008, (in press); DOI: 10.1016/ j.biombioe.2008.01.015

Chapter 5

Comparison of Fumonisin Concentrations in Kernels of Transgenic Bt Maize Hybrids and Nontransgenic Hybrids

Gary P. Munkvold[1], Richard L. Hellmich[2], and Larry G. Rice[3]

[1]Department of Plant Pathology,

[2]USDA-ARS Corn Insects and Crop Genetics Research Unit and Department of Entomology, Iowa State University, Ames 50011, and

[3]USDA-APHIS National Veterinary Services Laboratory, 1800 Dayton Rd., Ames, IA 50010

ABSTRACT

Munkvold, G. P., Hellmich, R. L., and Rice, L. G. 1999. Comparison of fumonisin concentrations in kernels of transgenic Bt maize hybrids and nontransgenic hybrids. Plant Dis. 83:130-138. Maize hybrids genetically engineered with genes from the bacterium *Bacillus thuringiensis* (*Bt* maize) express CryIA(b) and other Cry proteins that are toxic to certain insects, particularly the European corn borer (*Ostrinia nubilalis*). Maize kernel feeding by *O. nubilalis* often leads to infection by fungi in the genus *Fusarium*, including the fumonisin-producing species *F. verticillioides* and *F. proliferatum*. In field experiments in 1995, 1996, and 1997, transgenic maize hybrids and near-isogenic, nontransgenic hybrids were manually infested with neonatal European corn borer larvae. Manual infestation increased Fusarium ear rot severity and fumonisin concentrations in kernels of nontransgenic hybrids. Transgenic hybrids with kernel expression of CryIA(b) consistently experienced less insect feeding on kernels and less Fusarium ear rot than their nontransgenic counterparts. In manually infested treatments, these hybrids also exhibited lower concentrations of fumonisins in kernels compared with their nontransgenic counterparts. In manually infested treatments in 1995, mean fumonisin B1 concentrations were 8.8 μg/g in the nontransgenic hybrid and 6.7 or 3.0 μg/g in transgenic hybrids. In 1996, mean fumonisin B1 concentrations in manually infested treatments were 4.9 μg/g (range 2.3 to 8.8) for nontransgenic and 1.2 μg/g (range 1.0 to 1.3) for transgenic hybrids with kernel expression. Mean total fumonisin concentrations (fumonisin B1 + B2 + B3) were 7.0 μg/g (range 3.0 to 12.2) for nontransgenic and 1.7 μg/g (range 1.5 to 1.9) for transgenic hybrids with kernel expression. In 1997, mean fumonisin B1 concentrations in manually infested treatments were 11.8 μg/g (range 7.6 to 17.3) for nontransgenic and 1.3 μg/g (range 0.8 to 2.2) for transgenic hybrids with kernel expression of CryIA(b) or Cry9C. Mean total fumonisin concentrations were 16.5 μg/g (range 10.7 to 24.0) for nontransgenic and 2.1 μg/g (range 1.5 to 3.1) for transgenic hybrids with kernel expression. Transgenic hybrids that do not express CryIA(b) or Cry9C in kernels did not consistently have fumonisin concentrations different from the nontransgenic hybrids. Higher fumonisin concentrations in nontransgenic hybrids were associated with high European corn borer populations during the early reproductive stages of the maize plants. These results indicate that under some conditions, genetic engineering of maize for insect resistance may enhance its safety for animal and human consumption.

Fungi in the genus *Fusarium* are very common associates of maize (*Zea mays* L.) plants. These fungi can cause disease at all developmental stages of the plant, but also may infect plants without causing symptoms (24). Certain *Fusarium* species, particularly *F. verticillioides* (Sacc.) Nirenberg (synonym = *F. moniliforme* J. Sheld.) and *F. proliferatum* (T. Matsushima) Nirenberg, often infect maize plants through wounds caused by insects such as the European corn borer (*Ostrinia nubilalis*) (3,24). The same two species are capable of producing fumonisins, a class of mycotoxins that can cause fatal leucoencephalomalacia in horses, pulmonary edema in swine, and cancer in laboratory rats (13). Esophageal cancer in humans has been associated with consumption of maize with high concentrations of the fumonisins. The compounds are now classified as Class 2B, probable human carcinogens (10,20). Fumonisins can be found in maize grown throughout most of the world, and they are the most commonly occurring mycotoxins in maize grown in the midwestern United States (16,27). This class of toxins includes several homologs, but fumonisin B1 is typically found in the highest concentrations (16). Fumonisin-producing *Fusarium* species cause Fusarium ear or kernel rot, but fumonisins also can be detected in symptomless infected maize kernels (16).

The European corn borer (ECB) is a major pest of maize. The larvae of this insect tunnel into the stalks and ears of the plants, causing yield reduction and loss of grain quality (14). In the north central United States, two generations of ECB larvae typically occur per year; the first generation attacks plants during the mid- to late-vegetative stages (V8-V10 [22]), and the second generation attacks during the reproductive stages (R1 to maturity). Adult moths lay masses of eggs on leaf surfaces, and the larvae migrate to protected sites, where they feed on leaves, stalks, and ears (14). These larvae cause physical injury to the plants, where *Fusarium* infection can occur (3), and the larvae also can carry *Fusarium* spores from the leaf surface, transmitting them to the infection court (feeding site) (25). A new strategy for the management of *O. nubilalis* is the use of transgenic maize hybrids (*Bt* maize). Now commercially available, these hybrids contain *cry* genes originating from the bacterium *Bacillus thuringiensis*; gene expression results in the production of insecticidal crystalline proteins in plant tissue (8).

Currently available commercial hybrids express one of three Cry proteins: CryIA(b), CryIA(c), or Cry9C. Maize hybrids expressing Cry proteins have demonstrated a high level of resistance to feeding by ECB larvae (2,11). Expression of Cry proteins in specific maize tissues can be manipulated through the use of different gene promoters (11). Specific combinations of transgenes and gene promoters are often called "transformation events," and each combination must be approved separately by the U.S. Environmental Protection Agency. The use of these genetically engineered maize hybrids has been widely accepted in the United States but has met opposition elsewhere in the world because of concerns about food safety, environmental protection, and ethical issues.

Transgenic maize hybrids expressing CryIA(b) in kernels can experience less *Fusarium* infection in kernels because of the association between *Fusarium* infection and insect feeding (17). We hypothesized these hybrids also might contain lower concentrations of fumonisins than nontransgenic hybrids that are susceptible to the ECB. The objective of this study was to test this hypothesis through field experiments using near-isogenic maize hybrids with and without *cry* genes. *Fusarium* infection data from two of these experiments have been reported (17).

MATERIALS AND METHODS

Field experiments were conducted in 1995, 1996, and 1997. In each experiment, transgenic hybrids were compared with near-isogenic, nontransgenic maize hybrids lacking *cry* genes. Near-isogenic hybrids were supplied by cooperating seed companies and were either identical with (except for the *cry* genes) or very closely related to their corresponding transgenic hybrids. Experiments were arranged in randomized complete block designs in 1995 and 1996; in 1997, a split-plot design was used with hybrids as main plots and infestation treatments as subplots. Plot size differed among years: in 1995, plots were four rows by 9.15 m long (approximately 40 plants per row); in 1996, they were four rows by 6.1 m long (approximately 26 plants per row); and in 1997, main plots were four rows by 3.05 m long (approximately 15 plants per row) and subplots were single rows. Row spacing was 0.76 m. Only the middle two rows of the four-row plots were treated and analyzed. There were 3.05-m or 1.5-m fallow alleyways between replicate blocks.

The 1995 experiment, planted at the Iowa State University Woodruff Farm in Story County on 22 May, involved three maize hybrids (Table 1) and two *O. nubilalis* infestation treatments with six replicate blocks. The 1996 experiment, planted on 13 and 14 May at the ISU Johnson Farm in Story County, included eight hybrids (Table 1) and two *O. nubilalis* infestation treatments with five replicate blocks.

The 1997 experiment, planted on 20 May at the ISU Berkey Farm in Boone County, included 12 hybrids (Table 1) and four infestation treatments with 10 replicate blocks.

In 1995 and 1996, the two *O. nubilalis* infestation treatments were natural infestation and manual infestation. Laboratoryreared ECB larvae for manual infestation were obtained from the USDA Corn Insects Laboratory in Ames, Iowa. Larvae were premixed in ground corn and applied to the plants using a volumetric dispenser that dispenses 1.0-g aliquots of mixture containing 50 neonatal larvae. Manual infestation was conducted by dispensing 50 larvae into the whorl of each plant at growth stage V8-V10 (22) (first generation) and into the axil of the ear leaf at growth stage R1 (second generation).

Manual infestation was conducted to reduce plant-to-plant variability in insect feeding and to simulate a uniform infestation throughout the plots. This method and intensity for manual infestation are standard for European corn borer research (7). In 1997, the four infestation treatments were: natural infestation, manual infestation, *F. verticillioides*–contaminated first generation manual infestation (Manual- F1); and *F. verticillioides*–contaminated second-generation manual infestation (Manual-F2). Larvae for the Manual-F1 treatment were prepared by growing *F. verticillioides* strain M-8114 (Pennsylvania State University *Fusarium* Research Center, originally isolated in Iowa) on carnation- leaf agar (19) for 7 days, then scraping the contents of four CLA cultures into 500- ml bottles containing approximately 33,000 larvae. The contents were thoroughly mixed by hand before applying to the plants as already described. For the Manual-F2 treatment, five third-instar larvae were placed in 10-ml plastic beakers and lightly misted with a suspension of strain M-8114 conidia (approximately 105/ ml) in sterile water. One beaker was immediately placed into the axil of the ear shank of each of five plants per subplot at growth stage R1.

Surface contamination of the larvae in the F1 and F2 treatments with *F. verticillioides* was confirmed by placing 50 neonatal larvae or five third-instar larvae on Nash-Snyder Medium (19). *F. verticillioides* was isolated from 60 to 75% of neonatal and 100% of third-instar larvae, compared with 5 to 10% of the larvae that were not intentionally contaminated.

After the plants reached maturity (kernel moisture of approximately 20%) in 1995 and 1996, 10 arbitrarily selected ears per plot were visually evaluated in the field for Fusarium ear rot and insect feeding on kernels. In 1997, five ears per subplot were hand-harvested and evaluated in the laboratory.

For each ear, the numbers of kernels with symptoms of Fusarium ear rot and insect feeding were recorded. Incidences of Fusarium ear rot and insect feeding were calculated as the percentages of plants per plot with symptoms. Severity of Fusarium ear rot and insect feeding were calculated as the mean number of kernels per ear with symptoms. In 1995, insect feeding was recorded in a different way; a 0 to 3 scale was used in which 0 = no damage, 1 = £5 cm2 of damage per ear, 2 = >5 £ 10 cm2 of damage per ear, and 3 = >10 cm2 of damage per ear. Incidence and severity calculations were done as in the other years. The number of ECB larvae found on each plant also was recorded after splitting the stalks and ear shanks longitudinally and examining these tissues for larvae.

Table 1. Transformation events, Cry proteins, and maize hybrids represented in field experiments examining the effects of Cry protein expression on fumonisin concentrations in kernels

Year	Transfo-rmation	Cry protein	Cry protein epression in all plant tissues[t]	Hybrid
1995	MON810	CryIA(b)	Yes	Monsanto[u]802
	MON810	CryIA(b)	Yes	Monsanto[u]810
	None	None	....	B73×Mo17
1996	MON810	CryIA(b)	Yes	Monsanto[u]810

	None	None	.....	M73×Mo17
	BT11	CryIA(b)	Yes	Northup King[v]X6534CBR
	None	None	....	Northup King[v]N6800
	176	CryIA(b)	No	Mycogen[w]NG7059BT
	None	None	....	Mycogen[w]7050cb[x]
	176	CryIA(b)	No	Ciba[v]MAX454
	None	None	....	Ciba[v]4495
1997	MON810	CryIA(b)	Yes	Pioneer[x]hybrid34R06
	None			Pioneer[x]hybrid3489
	Bt11	CryIA(b)	Yes	Novartis[v]NK7070BT
	None			Novartis[v]NK7070
	176	CryIA(b)	YES	Mycogen[w]NK7059
	None			Mycogen[w]7059cb[x]
	176	CryIA(b)	No	Ciba[v]Max454
	None	None		Ciba[v]4495
	DBT418	CryIA(c)	Yes	DeKalb[y]580BT
	None	None		DeKalb[y]580
	CBH351	Cry9c	Yes	AgrEvo[z]Bt
	None	None	....	AgrEvo[z]non-Bt

[t] As indicated by supplier.

[u] Monsanto Co., St. Louis, MO.

[v] Novartis Seeds, Greensboro, NC.

[w] Mycogen Corp., San Diego, CA.

[x] Hybrid with naturally occurring resistance to European corn borer.

[y] DeKalb Plant Genetics, DeKalb, IL.

[z] Agrevo USA Co., Apple Valley, MN.

Table 2. P values for the effects of hybrid brand, presence of cry gene, and European corn borer (ECB) infestation on fumonisin concentrations in kernels for field experiment near Ames, Iowa, in 1996

Effect	Fumonisin B_1	Total fumonisins
Brand	0.0001	0.0001
cry gene	0.0002	0.0003
ECB infestationz	0.0001	0.0001
Block	0.0026	0.0014
Brand × *cry* gene	0.0065	0.0061
Brand × ECB	0.0798	0.0606
cry gene ×ECB	0.0008	0.0003
Brand × *cry* gene × ECB	0.0141	0.0112

In 1995 and 1996, 30 ears per plot were hand-harvested (including the ears evaluated for disease and insect feeding), and the kernels were removed by hand or by an electric sheller. Symptomless infection was determined by culturing 50 arbitrarily selected symptomless kernels (surface-disinfested for 3 min in a 0.5% solution of sodium hypochlorite) from each plot on Nash-Snyder medium (NSM) (19). In 1995 and 1996, *Fusarium* colonies were identified after 7 to 10 days by colony and conidial morphology (19). Colonies were not identified to species in 1995 and 1996 but were classified into groups of *Fusarium* spp., and the percentage of kernels (incidence) infected with each group was recorded. In 1997, after the five ears harvested from each subplot had been evaluated for ear rot and insect feeding severity, they were shelled with an electric sheller, and 25 symptomless kernels per subplot were cultured on NSM. From five infected kernels per subplot, a single colony was transferred from NSM to carnation leaf agar (CLA) and identified to species according to the keys and descriptions of Nelson et al. (19). Incidence of infection by each species was recorded.

In 1995 and 1996, a 0.6-kg sample was collected arbitrarily from the shelled kernels from each plot and ground in a Wiley mill with a 2-mm mesh screen. In 1997, after the 25-kernel sample was removed and cultured, the remainder of the kernels was ground for fumonisin analysis. The ground maize was thoroughly mixed, and 10-g samples were taken for analysis of fumonisin B1 in 1995 by the procedure of Hopmans and Murphy (9) and of fumonisins B_1, B_2, and B_3 (total fumonisins) in 1996 and 1997 by the procedure of Rice et al. (21). Briefly, 10-g samples were extracted with 50 ml of 50% acetonitrile (ACN) for at least 30 min. The supernatant was filtered through filter paper, and 2 ml of the filtered extract was diluted with 6 ml of 1% potassium chloride (KCl). The diluted sample was then cleaned up using a 500-mg C18 solid-phase extraction (SPE) column (Waters/Millipore, Milford, MA) by rinsing with 1 ml of 1% KCl and 3 ml of 15% ACN. The fumonisins were then eluted using 2 ml of 70% ACN. Fluorescent derivatives of the fumonisins were formed using *o*-phthaldialdehyde (OPA) and detected using a fluorescence detector (LC-240, Perkin-Elmer, Norwalk, CT) with an excitation wavelength of 335 nm and an emission wavelength of 440 nm. Derivatization was accomplished at room temperature using a 0.25-ml sample, 0.25 ml of borate buffer, (pH 8.3), 0.25 ml of OPA solution (15 mg of OPA and 20 µl of 2-mercaptoethanol in 10 ml of ACN), and 0.25 ml water. Reaction time was 10 min. Ten microliters of derivatized sample (0.5 mg equivalent) was injected into a 3 cm × 4.6 mm high speed, 3 µm packing material, C18 column (Perkin-Elmer) with quantitation made using external standards and peak height measurements. Fumonisin B1 and B2 standards were purchased from Sigma Chemical Co. (St. Louis, MO), and fumonisin B3 standard was obtained from Ron Plattner (USDA-ARS, NCAUR, Peoria, IL). Detection limits were 0.1 µg/g for fumonisin B1 and 0.5 µg/g for fumonisins B_2 and B_3.

Analysis of variance (SAS, Cary, NC, or SigmaStat, Jandel Scientific, San Diego, CA) was utilized to evaluate insect feeding severity, numbers of ECB larvae, Fusarium ear rot severity, incidence of symptomless infection, fumonisin B1 concentration, and total fumonisin concentration. A factorial analysis was used to evaluate the main effects and interactions of hybrid brand, presence of *cry* genes, and ECB infestation treatment. Multiple comparison tests were performed according to the Student-Newman- Keuls test on treatment means if there were interactions. A square-root transformation was used to obtain a normal distribution for the fumonisin data. Linear correlation coefficients were calculated for the relationships among severity of Fusarium ear rot, insect feeding severity, and fumonisin concentration (SigmaStat), using data from individual experimental units.

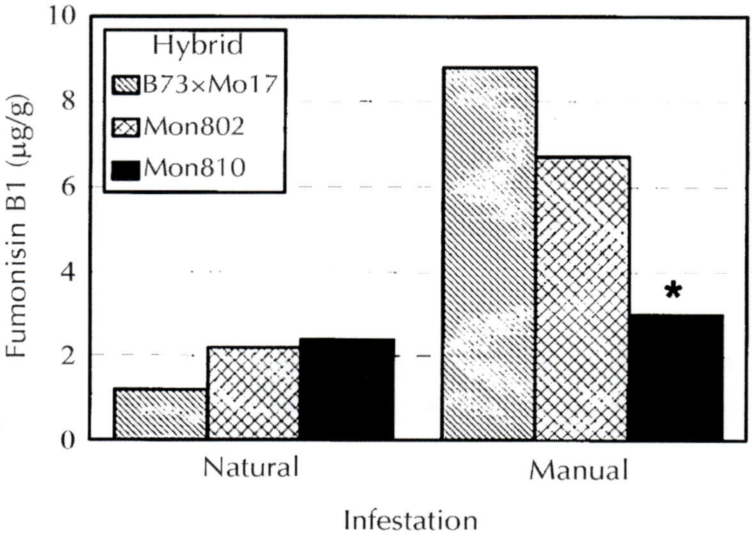

Figure 1: Fumonisin B1 concentrations in kernels of transgenic *Bt* maize hybrids (Mon802 and Mon810) and a near-isogenic standard hybrid (B73×Mo17) in 1995. Manually infested plants were infested with 50 neonatal European corn borer (*Ostrinia nubilalis*) larvae at growth stages V8 to V10 and R1. * indicates a significant difference (*P* £ 0.05) between standard and transgenic hybrid.

Table 3: *P* values for the effects of hybrid brand, presence of *cry* gene, and European corn borer (ECB) infestation on Fusarium kernel rot incidence, Fusarium ear rot severity, and incidence of symptomless kernel infection by *Fusarium* spp. in a field experiment in Boone County, Iowa, in 1997

Effect			Fusarium ear rot			Symptomless *Fusarium* infection`			Fumonisin concentration	
	Incidence"	Severity'	Incidence"	Severity'	Total	*F vent.*	*F.proL*	*F subg.*	FB_1	Total
Brand	0.0001	0.0001	0.0001	0.0001	0.0001	0.0001	0.0003	0.0189	0.5145	0.5684
cry genet	0.0001	0.0001	0.0001	0.0001	0.0001	0.0001	0.1188	0.2815	0.0001	0.0001
ECB infestation 4	0.1300	0.0989	0.0040	0.0019	0.0001	0.0001	0.8722	0.0001	0.0001	0.0001
Block	0.1149	0.0016	0.2693	0.0079	0.0003	0.0001	0.7183	0.1943	0.0152	0.0128
Brand x *cry* gene	0.0001	0.0001	0.0001	0.0001	0.0001	0.0001	0.0120	0.0908	0.0001	0.0001
Brand x ECB	0.8711	0.2805	0.1352	0.3543	0.0051	0.0077	0.1536	0.3557	0.9550	0.9419
cry gene x ECB	0.3668	0.2739	0.0322	0.0482	0.4020	0.5560	0.4382	0.2786	0.0338	0.0345
Brand x *cry* gene x ECB	0.8586	0.2812	0.4718	0.2728	0.4261	0.4578	0.1545	0.5433	0.7115	0.6732

v Measured as the percentage of infected kernels, based on isolation from 10 replications of 25 kernels each. *F. verticillioides*, *F. proliferatum*, and *F. subglutinans*.

w Measured as the percentage of plants with symptoms, based on 10 replications of five plants each.

x Measured as the mean number of kernels with symptoms per ear, based on 10 replications of five plants each.

y Effect of presence of cry gene.

z There were four infestation treatments: natural ECB infestation, manual infestation, or manual infestation with *F. verticillioides* contamination at growth stage V8 to V10 or R1.

RESULTS

Hybrid brand, presence of *cry* genes, and insect infestation treatments had effects on fumonisin B1 and total fumonisin concentrations, and the significance of these effects differed among years. In 1995, when only one brand was included, the main effect of hybrid on fumonisin B1 concentration was not significant ($P = 0.240$), but infestation had a significant effect ($P = 0.032$) and there was a significant interaction ($P = 0.044$). In 1996, effects of hybrid brand, presence of *cry* genes, and infestation treatment were significant, but there were significant two-way and three-way interactions (Table 2). In 1997, the significant main effects were those of *cry* gene presence and infestation treatment, and there were significant two-way interactions (Table 3). Fumonisin concentrations were significantly lower in some transgenic hybrids compared with their nontransgenic counterparts, and fumonisin concentrations were enhanced when maize plants were manually infested with ECB larvae (Figs. 1 to 3). Significant differences among hybrids were more evident under manual infestation. In 1997, the fumonisin concentrations were higher than in the other 2 years, and the differences between transgenic and nontransgenic hybrids were most dramatic. Transgenic hybrids in some cases had concentrations less than 10% of their nontransgenic counterparts. The Manual- F1 treatment resulted in the highest fumonisin concentrations in 1997, with a mean of 17.3 µg/g (total fumonisins), compared with 12.1 µg/g in the Manual-F2 treatment, 11.3 µg/g in the manually infested treatment, and 7.8 µg/g in the naturally infested treatment. Fumonisin B1 accounted for approximately 70% of the total fumonisins, and ANOVA on total fumonisins yielded results almost identical to those for fumonisin B1.

In all 3 years, presence of *cry* genes had a significant (*P* £ 0.0001) effect on insect feeding severity on ears, with lower severity in transgenic hybrids than in nontransgenic hybrids (Tables 4 and 5). In 1996, the effects of hybrid brand and the presence of *cry* genes were significant (*P* £ 0.0001), but there also were significant two-way and three-way interactions (*P* £ 0.05). Insect infestation had a significant effect on feeding severity in 1996 (*P* £ 0.0001) but not in 1995 and 1997 ($P = 0.8235$ in 1995 and $P = 0.0989$ in 1997). Significantly more larvae were found in naturally infested plants than in manually infested plants ($P = 0.0043$ in 1995 and $P = 0.0277$ in 1996) (Tables 4 and 5). Numbers of larvae were affected significantly by hybrid brand and presence of *cry* genes (transgenic hybrids had fewer larvae than nontransgenic hybrids) (Tables 4 and 5), and there was a significant ($P = 0.001$) interaction between these two factors. Fusarium ear rot severity was affected by hybrid brand, presence of *cry* genes, and ECB infestation (*P*£ 0.0001), but there were significant interactions among these factors (17; Table 6). Generally, symptoms were less for transgenic hybrids than for nontransgenic, and less for natural infestation than for manual infestation, but these effects depended on the hybrid brand. In 1997, the second-generation manual infestation treatment that included *F. verticillioides* contamination (Manual-F2) had the highest ear rot severity (Table 6).

Incidence of symptomless kernel infection by *Fusarium* spp. was significantly affected by hybrid brand, presence of *cry* genes, and infestation treatment; there also were significant two-way interactions (17; Table 6). Transgenic hybrids had lower incidence of symptomless infection (43.4% in 1997) than nontransgenic hybrids (53.5% in 1997), but the difference depended on hybrid brand and infestation treatment. In 1997, the Manual-F1 treatment resulted in the highest incidence of symptomless infection in 11 of 12 hybrids (Table 6). Three species, *F. verticillioides, F. proliferatum,* and *F. subglutinans*, accounted for 99.0% of the *Fusarium* colonies identified from symptomless infected kernels. *F. verticillioides* comprised the majority (81.4%) of the *Fusarium* colonies identified from all treatments. Analysis of variance on individual species indicated that incidence of infection by *F. verticillioides* was higher in the manual ECB infestation treatments than in the natural infestation treatment, but incidence of the other two species was either the same or lower in the manual infestations than in the natural infestation. In 1996 and 1997, insect feeding severity, Fusarium ear rot severity, and incidence of symptomless infection were significantly correlated with fumonisin B1 concentration (Fig. 4). In 1995, fumonisin B1 concentrations were significantly correlated only with incidence of symptomless infection.

DISCUSSION

Under some conditions, transgenic maize hybrids expressing *cry* genes from *B. thuringiensis* can contain lower concentrations of fumonisins than do nontransgenic hybrids. These lower concentrations were associated with earlier-reported reductions in Fusarium ear rot and/or symptomless infection (17) that, in turn, can be attributed to reduced insect feeding on kernels of transgenic hybrids. This finding indicates that the safety of maize for human and animal consumption may be enhanced by genetic engineering for insect resistance. The magnitude of the differences in fumonisin concentrations between transgenic and nontransgenic hybrids was sufficient to affect the toxicity of these maize kernels to horses (15) and to human cell cultures (1, 23, 26). The fumonisin concentrations that occurred in some nontransgenic hybrids were high enough to be considered unsafe for some animal species and inadvisable for human consumption. The American Association of Veterinary Laboratory Diagnosticians (AAVLD) has recommended that feeds for horses contain less than 5 μg of fumonisins per g and that feeds for swine contain less than 10 μg of fumonisins per g (15). One or both of these criteria were exceeded by nontransgenic hybrids in all 3 years. There are no widely accepted standards regarding safe fumonisin concentrations for human consumption. The hybrid brand × *cry* gene interaction is important in the analysis of our results because it indicates whether Bt transformation events differ in their effects on the dependent variables that we measured. The transgenic hybrids included in the 1996 and 1997 experiments are the products of different transformation events, so they differ either in the gene promoter or the Cry protein. Directly comparing data from different transgenic hybrids is problematic because the background *Fusarium* susceptibility of the hybrids differs and the effects of background genetics cannot be separated from the effects of the transgenes. By examining the hybrid brand × *cry* gene interactions, we can assess how different transformation events affected the Fusarium ear rot, *Fusarium* infection, and fumonisin concentrations of their respective background hybrids.

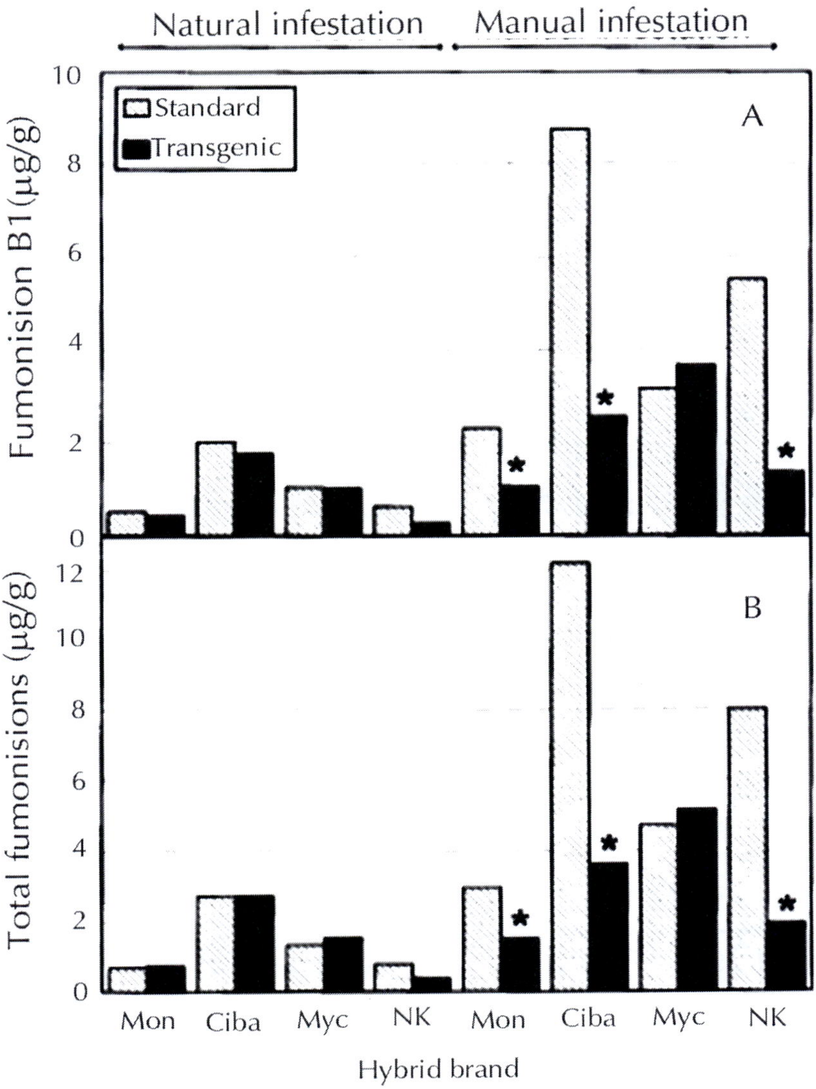

Figure 2: Fumonisin B1 **(A)** and total fumonisin **(B)** concentrations in kernels of transgenic *Bt* maize hybrids and near-isogenic standard hybrids in 1996. Transgenic hybrids are paired with their nearisogenic counterparts. Manually infested plants were infested with 50 neonatal European corn borer (*Ostrinia nubilalis*) larvae at growth stages V8 to V10 and R1. Hybrids are identified by brand: Mon (Monsanto), Ciba (Novartis), Myc (Mycogen), NK (Novartis). * indicates a significant difference (*P* £ 0.05) between standard and transgenic hybrids.

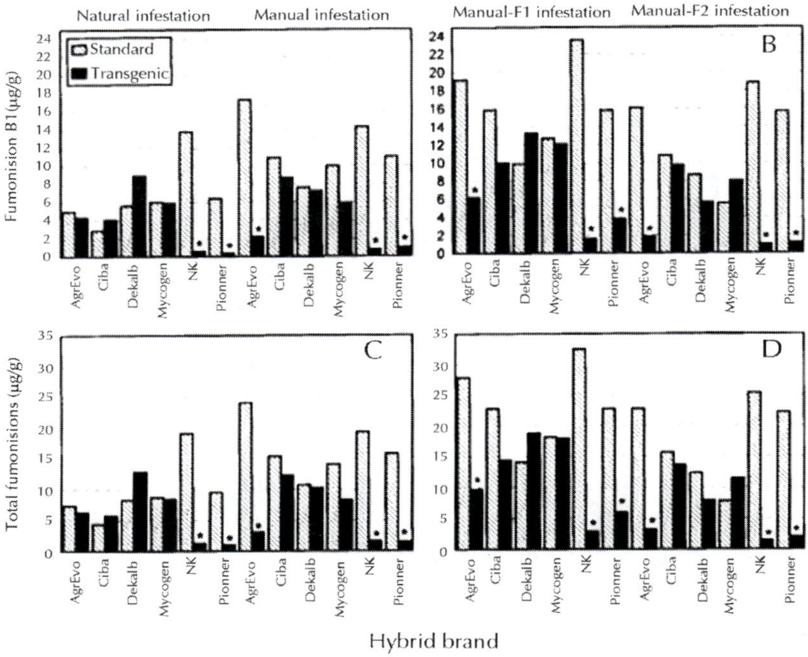

Figure 3: Fumonisin B1 **(A and B)** and total fumonisin **(C and D)** concentrations in kernels of transgenic *Bt* maize hybrids and near-isogenic standard hybrids in 1997. Transgenic hybrids are paired with their near-isogenic counterparts. Manually infested plants were infested with 50 neonatal European corn borer (*Ostrinia nubilalis*) larvae at growth stages V8 to V10 and R1; plants in the manual-F1 treatment were infested with 50 *Fusarium verticillioides*–contaminated neonatal larvae at stages V8 to V10; and plants in the manual-F2 treatment were infested with five *F. verticillioides*–contaminated thirdinstar larvae at stage R1. Hybrids are identified by brand. * indicates a significant difference (*P* £ 0.05) between standard and transgenic hybrids.

These interactions were often significant, indicating that the MON810 and BT11 transformation events led to significantly greater reductions in Fusarium ear rot severity, incidence of symptomless infection, and fumonisin concentrations than did the 176 event. The differential effects of the transformation events can be attributed to differences in *cry* gene expression. Maize hybrids with the MON810 or BT11 transformation types utilize a CaMV 35S gene promoter (2) that results in seasonlong expression of CryIA(b) in all plant tissues; whereas transformation 176 utilizes a combination of two maize-derived, tissue-specific promoters: a phosphoenolpyruvate carboxylase (PEPC) promoter that results in gene expression only in green plant tissues, and a pollen-specific promoter (11). Hybrids expressing CryIA(b) in kernels were better protected against *Fusarium* infection than those expressing the protein only in green tissues and pollen. This trend was evident but not universal. Hybrids with the MON810 and BT11 transformation events consistently demonstrated significant reductions in *Fusarium* infection and fumonisin concentration in the manual infestation treatments.

Although hybrid Mon802 also expresses CryIA(b) in kernels, it did not exhibit reduced fumonisin concentration. This hybrid expresses lower concentrations of CryIA(b) than Mon810 (5), and the lower expression in this hybrid may not have controlled insect feeding on kernels sufficiently to affect fumonisin concentrations. Insect feeding severity for this hybrid was higher than, but not significantly different from, insect feeding severity for Mon810. Therefore, the difference in fumonisin B1 concentration between these two hybrids cannot be explained completely with the available data. Hybrids with the 176 transformation event do not have kernel expression of CryIA(b) and did not consistently demonstrate effects on Fusarium ear rot or fumonisins. Ciba hybrid Max454 demonstrated significantly less Fusarium ear rot in 1996 and 1997 and lower fumonisin concentrations in 1996 than its nontransgenic counterpart, 4494, but the Mycogen transgenic hybrid did not exhibit any significant reductions in any of the variables in the manual infestation treatment. This result might have been due to differences in the genetic backgrounds of the hybrids; the nontransgenic Mycogen hybrid is known to have native resistance to *O. nubilalis*. The DBT418 transformation event demonstrated results similar to the 176 event, and the CBH351 event demonstrated results similar to the MON810 event, but we have only 1 year of data for DBT418 and CBH351.

Table 4: Insect feeding severity on maize ears and European corn borer (ECB) larvae per plant in transgenic (Bt) and non-transgenic hybrids in field experiments conducted in 1995 and 1996v

Year	Hybrid		Insect feeding severity*			ECB larvae ;plant"	
		Natural	Manual[y]	Mean	Natural	Manual[y]	Mean
1995	B73 x Mo17	0.7 a	0.7 a	0.7	1.17 a	0.42b	0.79
	Mon 802*[z]	02b	0.3 b	0.2	0.02 c	0.00c	0.01
	Mon 810*	0.1 b	0.1 b	0.1	0.00 c	0.00c	0.00
	Means	0.3	0.4	0.3	0.40	0.14	0.27
1996	Mon 810*	0.5ef	0.2f	0.3	0.00f	0.00 f	0.00
	B73 x Mo17	2.2def	2.3 def	2.2	1.02a	0.86 abc	0.96
	Ciba Max454	3.0cde		2.6	0.16ef	0.10 f	0.13
	Ciba 4494	5.1 bc	12.3 a	8.7	0.94ab	0.80abc	0.87
	MycogenNG7-059BT*	3.3 cd	7.1 b	5.2	0.52cde	0.20def	0.36
	Mycogen 705cb	6.3 b	7.6 b	6.9	0.78abc	0.52cde	0.66
	Northnip King X6534CBR*	0.8 def	0.4 ef	0.6	0.00f	0.00f	0.00
	Northrup King N6800	2.0def	7.0 b	4.5	0.58k	0.54 cd	0.56
	Means	2.9	4.9	3.9	0.50	0.38	0.44

v Near-isogenic hybrids are paired in the table. Values are means for six replications of 10 ears in 1995 and five replications of 10 ears in 1996. Values within a year followed by the same letter are not significantly different (Student-Newman-Keuls test, a = 0.05).

w Insect feeding severity was assessed on a 0 to 3 scale in 1995 and as the mean number of damaged kernels per plant in 1996 (10 plants per plot).

x Numbers of live and dead ECB larvae in the ears, stalks, and shanks were recorded at harvest maturity (10 plants per plot).

y Manual infestation consisted of 50 neonatal ECB larvae placed in the whorl at stage V8 to V10 and in the axil of the ear leaf at stage R1.

z * = Bt hybrids.

Table 5: Insect feeding severity on maize ears in transgenic (Bt) and nontransgenic hybrids in a field experiment conducted in 1997u

Year	Hybrid			Insect feeding severity`'		
		Natural	Manually	Manual-Fl=	Manual-Fr	Mean
1997	Pioneer hybrid 34R06•1	0.5	0.7	1.7	1.3	1.0 e
	Pioneer hybrid 3489	12.8	15.7	16.0	22.6	16.8 be
	Novartis NK 7070BT*	0.7	0.3	0.7	1.3	0.7 e
	Novartis NK 7070	17.1	20.8	19.9	20.6	19.6 b
	Mycogen NG 7059*	11.7	13.6	13.4	12.6	12.8 cd
	Mycogen 7059cb	14.9	15.1	16.3	11.6	14.5 cd
	Ciba Max 454*	9.9	11.4	8.7	13.4	10.8 d
	Ciba 4494	15.3	22.1	17.6	20.0	18.7 b
	DeKalb 58OBT*	9.7	11.1	13.2	9.3	10.8 d
	DeKalb 580	18.9	15.8	13.1	14.5	15.6 be
	AgrEvo Bt*	4.4	3.1	1.9	2.3	2.9 e
	AgrEvo	28.3	34.2	26.1	36.7	31.3 a
	Means	12.0	13.7	12.4	13.9	13.0

u Near-isogenic hybrids are paired in the table. Values are means for 10 replications of five ears. Values followed by the same letter are not significantlydifferent (Student-Newman-Keuls test, a = 0.05).

v Insect feeding severity was assessed as the mean number of damaged kernels per plant in 1997 (five ears per plot).

w Manual infestation consisted of 50 neonatal European corn borer (ECB) larvae placed in the whorl at stage V8 to V10 and in the axil of the ear leaf atstage R1.

x Manual-F1 treatment consisted of 50 neonatal ECB larvae contaminated with *Fusarium verticillioides* conidia and placed in the whorl at stage V8 toV10.

y Manual-F2 treatment consisted of five third-instar ECB larvae contaminated with *F. verticillioides* conidia and placed in the ear shank axil at stage R1.

z * = Bt hybrids.

Fumonisin concentrations were significantly correlated with insect feeding severity, Fusarium ear rot, and symptomless *Fusarium* infection in 1996 and 1997, but only with symptomless infection in 1995. Symptomless infection occurs by a number of pathways that may or may not be related to insect feeding (18). Typically, fumonisin concentrations are higher in symptomatic kernels (6), and Fusarium ear rot severity is enhanced by insect feeding (18).

Table 6: Fusarium ear rot severity and symptomless kernel infection in Bt and non-Bt hybrids subjected to manual and natural infestations of European corn borer (ECB) in Boone County, Iowa, in 1997u

Brand	Hybrid	Fusarlum ear rot severity (kernels/ear)					Symptomless infection (% of kernels)				
		Manual[y]	Manual-F1[w]	Manual-F2[x]	Natural[y]	Mean	Manual[v]	Manual-Fl[w]	Manual-F-F2[x]	Natural[y]	Mean
AgrEvo	Bin	2.8 c	2.5 d	3.2 c	4.8 c	3.3	25.9 d	56.4 a	30.4 c	19.6 d	33.1
AgrEvo	non-Bt	38.4 a	23.1 ab	31.9 a	21.5 a	28.7	43.5 bcd	64.4 a	48.4 ab	30.2 b-e	46.6
Ciba	Max454*	10.1 be	10.6 e	13.9 b	9.2 be	11.0	67.8 ab	64.4 a	60.0 ab	46.7 ab	59.7
Ciba	4454	25.7 b	17.6 be	26.6 ab	14.1 ab	21.0	71.6 a	72.8 a	66.0 a	49.6 a	65.0
De/alb	DK58081'*	17.9 b	19.2 abc	19.1 ab	15.7 ab	18.0	58.0 abc	70.0 a	58.4 ab	44.8 ab	57.8
DeKalb	DK580	19.8 b	18.5 abc	18.9 ab	21.5 a	19.7	59.1 abc	73.0 a	61.1 ab	39.2 abc	58.1
Mycogen	NG7059*	13.7 be	15.9 be	15.4 b	13.6 ab	14.7	54.8 abc	65.3 a	54.9 ab	37.8 a-d	53.2
Mycogen	7050cb	17.1 b	18.9 abc	16.2 b	16.2 ab	17.1	40.4 cd	60.4 a	41.2 be	28.7 bcd	42.7
/4K	M7070131'	2.1c	2.1 d	2.6c	1.3c	2.0	26.4 d	39.6 b	29.6c	19.2d	28.7
NK	NIC7070	25.1 b	26.6 a	27.2 ab	20.3 a	24.8	59.6 abc	67.6 a	57.2 ab	53.2 a	59.4
Pioneer	34R06'	1.7c	2.3d	2.5c	0.5c	1.8	28.4d	57.2a	12.0d	13.2e	27.7
Pioneer	3489	17.4 b	15.5 k	26.5 ab	15.6 ab	18.8	56.9 abc	66.2 a	52.0 ab	24.0 cd	49.8
Means		16.0	14.4	17.0	12.9	15.1	49.3	63.1	47.6	33.7	48.4

u Values in a column followed by the same letter are not significantly different (Student-Newman-Keuls test, a = 0.05).

v Manual infestation treatment consisted of 50 neonatal ECB larvae placed in the whorl at stage V8 to V10 and in the axil of the ear leaf at stage R1.

w Manual-F1 treatment consisted of 50 neonatal ECB larvae contaminated with *Fusarium verticillioides* conidia and placed in the whorl at stage V8 toV10.

x Manual-F2 treatment consisted of five third-instar ECB larvae contaminated with *F. verticillioides* conidia and placed in the ear shank axil at stage R1.

y The natural infestation treatment consisted of the endemic ECB population.

z * = Bt hybrids.

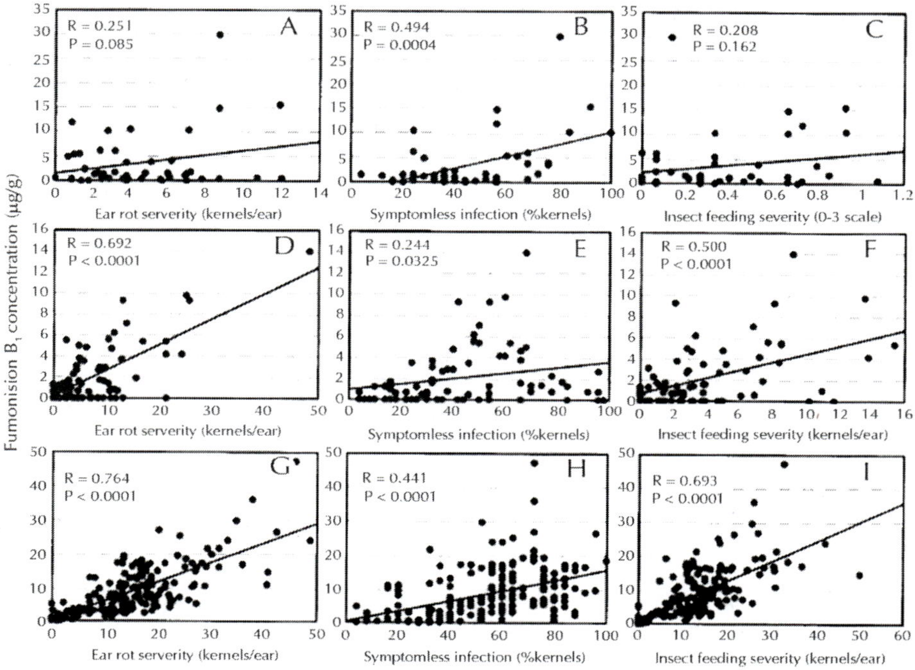

Figure 4: Linear correlations of Fusarium ear rot severity, symptomless *Fusarium* infection, and insect feeding severity with fumonisin B1 concentrations in 1995 **(A to C)**, 1996 **(D to F)**, and 1997 **(G to I)**. Correlations were calculated on individual experimental units.

The highest fumonisin concentrations occurred in 1997, the year in which insect feeding severity and Fusarium ear rot severity were highest. The 1996 and 1997 correlations are consistent with these relationships, but the 1995 correlations are less so. The lack of correlation between fumonisin concentrations and ear rot severity in 1995 is a result of the low fumonisin concentration in hybrid B73×Mo17 in the naturally infested compared with the manually infested treatment. Although manual infestation did not affect ear rot severity, it strongly affected fumonisin concentration. Sampling may also have affected correlation coefficients. In 1995 and 1996, Fusarium ear rot and insect feeding severity were assessed in the field on 10 ears per plot, but symptomless infection and fumonisin concentrations were analyzed on a subsample of grain harvested from each entire plot. In 1997, all assessments and analyses were performed on the same five ears per plot. As a result, correlation coefficients tended to be higher in 1997 than in 1996.

Our results also indicate that fumonisin concentrations in maize kernels can be affected by ECB infestation. Differences in Fusarium ear rot and fumonisin concentrations between transgenic and nontransgenic hybrids were evident primarily in manually infested treatments. The effect may be related to larval numbers, placement, and/or timing of infestation. Manual infestation with ECB larvae was performed in order to reduce variability in insect feeding among the plants and ensure that some viable larvae occurred on each plant. The number of larvae used (50) is a standard and realistic population for manual infestation (6,14). Natural mortality of ECB larvae hatching on maize leaf surfaces is about 70% in the first instar, and 50 to 80% of the surviving larvae die before tunneling into the plant (14). The number of larvae likely to survive in the manual infestation was therefore similar to that which would occur if one egg mass were laid per plant. Manual infestation did not increase the total number of larvae recovered from plants at maturity in 1995 or 1996, nor did it affect insect feeding severity on kernels in 1995 or 1997, indicating that the number of larvae per plant was not unrealistic in the manual infestation treatment. Nevertheless, this treatment did significantly affect Fusarium ear rot and fumonisin concentrations in some hybrids. This is probably more related to the timing and placement of the infestation than to the total number of larvae. Manual second-generation infestation occurred at stage R1, and larvae were placed in the ear-leaf axil; both timing and placement were optimal for the larvae to damage the kernels at an early plant reproductive stage, allowing the maximum time for fumonisin accumulation before harvest. In 1997, the manual infestation treatments that included *F. verticillioides* inoculum resulted in the highest Fusarium ear rot severity and fumonisin concentrations, as expected.

Other mycotoxins in maize also may be affected by insect activity. Most important among these are the aflatoxins, produced primarily by *Aspergillus flavus* (12). Seed of transgenic *Bt* cotton sometimes have lower aflatoxin concentrations than seed of nontransgenic cotton (4), but there are no reports demonstrating a similar effect in corn kernels. This possibility warrants further study.

Acknowledgments

Journal Paper J-17323 of the Iowa Agriculture and Home Economics Experiment Station, Ames, Project 3260, supported by Hatch Act and

State of Iowa funds. We thank Patricia A. Murphy, Iowa State University Department of Food Science and Human Nutrition, for cooperation in performing fumonisin analysis in 1995, Gemma Castellá, Autonomous University of Barcelona, for assistance with *Fusarium* identification in 1997, and John Shriver for technical assistance. This research was funded, in part, by the Monsanto Co., St. Louis, MO. This is a joint contribution from USDA, Agricultural Research Service, and Iowa State University. Product names are necessary to report factually on available data; however, neither the USDA nor Iowa State University guarantees or warrants the standard of the product, and the use of the names implies no approval of the product to the exclusion of others that may be suitable.

REFERENCES

1. Abbas, H. K., Tanaka, T., and Shier, W. T. 1995. Biological activities of synthetic analogues of *Alternaria alternata* toxin (AALtoxin) and fumonisin in plant and mammalian cell cultures. Phytochemistry 40:1681-1689.
2. Armstrong, C. L., Parker, G. B., Pershing, J. C., Brown, S. M., Sanders, P. R., Duncan, D. R., Stone, T., Dean, D. A., DeBoer, D. L., Hart, J., Howe, A. R., Morrish, F. M., Pajeau, M. E., Petersen, W. L., Reich, B. J., Rodriguez, R., Santino, C. G., Sato, S. J., Schuler, W., Sims, S. R., Stehling, S., Tarochione, L. J., and Fromm, M. E. 1995. Field evaluation of European corn borer control in progeny of 173 transgenic corn events expressing an insecticidal protein from *Bacillus thuringiensis*. Crop Sci. 35:550-557.
3. Christensen, J. J., and Schneider, C. L. 1950. European corn borer (*Pyrausta nubilalis* Hbn.) in relation to shank, stalk, and ear rots of corn. Phytopathology 40:284-291.
4. Cotty, P. J., Howell, D. R., Bock, C., and Tellez, A. 1997. Aflatoxin contamination of commercially grown transgenic BT cottonseed. Pages 108-110 in: Proc. Beltwide Cotton Prod. Res. Conf. National Cotton Council of America, Memphis, TN.
5. Davis, P. M., and Coleman, S. B. 1997. European corn borer (Lepidoptera: Pyralidae) feeding behavior and survival on transgenic corn containing CryIA(b) protein from *Bacillus thuringiensis*. J. Kans. Entomol. Soc. 70:31-38.

6. Desjardins, A. E., and Plattner, R. D. 1998. Distribution of fumonisins in symptomatic and symptomless kernels of maize. Plant Dis. 82:953-958.

7. Guthrie, W. D., and Barry, B. D. 1987. Methodologies used for screening and determining resistance in maize to the European corn borer. Pages 122-129 in: Toward insect resistant maize for the Third World: Proc. Int. Sympos. Methodol. Devel. Host Plant Resist. Maize Insects. CIMMYT, Mexico, D. F., Mexico.

8. Hofte, H., and Whiteley, H. R. 1989. Insecticidal crystal proteins of *Bacillus thuringiensis*. Microbiol. Rev. 53:242-255.

9. Hopmans, E. C., and Murphy, P. A. 1993. Detection of fumonisins B1, B2, and B3 and hydrolyzed fumonisin B1 in corn-containing foods. J. Agric. Food Chem. 41:1655-1658.

10. International Agency for Research on Cancer (IARC). 1993. Monograph on the Evaluation of Carcinogenic Risk to Humans, 56, 445.

11. Koziel, M. G., Beland, G. L., Bowman, C., Carozzi, N. B., Crenshaw, R., Crossland, L., Dawson, J., Desai, N., Hill, M., Kadwell, S., Launis, K., Lewis, K., Maddox, D., McPherson, K., Meghji, M. R., Merlin, E., Rhodes, R., Warren, G. W., Wright, M., and Evola, S. V. 1993. Field performance of elite transgenic maize plants expressing an insecticidal protein derived from *Bacillus thuringiensis*. Bio/Technology 11:194-200.

12. Lillehoj, E. B., Fennell, D. I., Kwolek, W. F., Adams, G. L., Zuber, M. S., Horner, E. S., Widstrom, N. W., Warren, H., Guthrie, W. D., Sauer, D. B., Findley, W. R., Manwiller, A., Josephson, L. M., and Bockholt, A. J. 1978. Aflatoxin contamination of corn before harvest: *Aspergillus flavus* association with insects collected from developing ears. Crop Sci. 18:921-924.

13. Marasas, W. F. O. 1995. Fumonisins: Their implications for human and animal health. Nat. Toxins 3:193-198.

14. Mason, C. E., Rice, M. E., Calvin, D. D., Van Duyn, J. W., Showers, W. B., Hutchison, W. D., Witkowski, J. F., Higgins, R. A., Onstad, D. W., and Dively, G. P. 1996. European corn borer ecology and management. NC Regional Extension Publ. 327. Iowa State University, Ames.

15. Miller, M. A., Honstead, J. P., and Lovell, R. A. 1996. Regulatory aspects of fumonisin with respect to animal feeds. Pages 363-368

in: Fumonisins in Food. L. Jackson, J. DeVries, and L. Bullerman, eds. Plenum Press, New York.

16. Munkvold, G. P., and Desjardins, A. E. 1997. Fumonisins in maize: Can we reduce their occurrence? Plant Dis. 81:556-565.
17. Munkvold, G. P., Hellmich, R. L., and Showers, W. B. 1997. Reduced Fusarium ear rot and symptomless infection in kernels of maize genetically engineered for European corn borer resistance. Phytopathology 87:1071-1077.
18. Munkvold, G. P., McGee, D. C., and Carlton, W. M. 1997. Importance of different pathways for maize kernel infection by *Fusarium moniliforme*. Phytopathology 87:209-217.
19. Nelson, P. E., Toussoun, T. A., and Marasas, W. F. O. 1983. Fusarium Species: An Illustrated Manual for Identification. Pennsylvania State University, University Park.
20. Rheeder, J. P., Marasas, W. F. O., Thiel, P. G., Sydenham, E. W., Shephard, G. S., and van Schalkwyk, D. J. 1992. *Fusarium moniliforme* and fumonisins in corn in relation to human esophageal cancer in Transkei. Phytopathology 82:353-357.
21. Rice, L. G., Ross, P. F., DeJong, J., Plattner, R. D., and Coats, J. R. 1995. Evaluation of a liquid chromatographic method for the determination of fumonisins in corn, poultry feed, and *Fusarium* culture material. J. AOAC Int. 78:1002-1009.
22. Ritchie, S. W., Hanway, J. J., and Benson, G. O. 1992. How a corn plant develops. Ia. State Univ. Ext. Spec. Rep. 48.
23. Shier, W. T., Abbas, W. K., and Mirocha, C. J. 1991. Toxicity of the mycotoxins fumonisin B1 and B2 and *Alternaria alternata* f. sp. *lycopersici* toxin (AAL) in cultured mammalian cells. Mycopathologia 116:97-104.
24. Smith, D. R., and White, D. G. 1988. Corn diseases. Pages 687-766 in: Corn and Corn Improvement, 3rd ed. G. F. Sprague and J. W. Dudley, eds. American Society of Agronomy, Madison, WI.
25. Sobek, E. A. 1996. Associations between corn insects and symptomatic and asymptomatic corn kernel infection by *Fusarium moniliforme*. M.S. thesis. Iowa State University, Ames.
26. Tolleson, W. H., Melchior, W. B., Jr., Morris, S. M., McGarrity, L. J., Domon, O. E., Muskhelishrili, L., James, S. J., and Howard, P.

C. 1996. Apoptotic and antiproliferative effects of fumonisin B1 in human keratinocytes, fibroblasts, esophageal epithelial cells and hepatoma cells. Carcinogenesis 17:239-249.

27. USDA-APHIS. 1996. Veterinary Services Information Sheet 3374.

Chapter 6

Crop Management Techniques to Enhance Harvest Index in Rice

Jianchang Yang[1], and Jianhua Zhang[2]

[1]Key Laboratory of Crop Genetics and Physiology of Jiangsu Province, Yangzhou University, Yangzhou, Jiangsu, China

[2]Department of Biology, Hong Kong Baptist University, Hong Kong, China

ABSTRACT

A major challenge in rice (*Oryza sativa* L.) production is to enhance water use efficiency (*WUE*) and maintain or even increase grain yield. *WUE*, if defined as the biomass accumulation over water consumed, may be fairly constant for a given species in given climate. *WUE* can be enhanced by less irrigation. However, such enhancement is largely a trade-off against lower biomass production. If *WUE* is defined as the grain production per unit amount of water irrigated, it would be possible to increase *WUE* without compromising grain yield through the manipulation of harvest index. Harvest index has been shown to be a variable factor in crop production, and in many situations, it is closely associated with *WUE* and grain yield in cereals. Taking rice as an example, this paper discussed crop management techniques that can enhance harvest index. Several practices such as post-anthesis controlled soil drying, alternate wetting and moderate soil drying regimes during the whole growing season, and non-flooded straw mulching cultivation, could substantially enhance *WUE* and maintain or even increase grain yield of rice, mainly via improved canopy structure, source activity, sink strength, and enhanced remobilization of pre-stored carbon reserves from vegetative tissues to grains. All the work has proved that a proper crop management holds great promise to enhance harvest index and, consequently, achieve the dual goal of increasing grain production and saving water.

INTRODUCTION

Global agriculture in the 21st century faces two major challenges. Total food production needs to increase to feed a still-growing world population and this increase needs to be accomplished under increasing scarcity of water resources (Bouman, 2007). Rice (*Oryza sativa* L.) is one of the most important crops in the world and is the foremost staple food in Asia, providing 35–60% of the dietary calories consumed by nearly three billion people (Fageria, 2003). By the year 2025, it will be necessary to produce about 60% more rice than is currently being produced to meet the food needs of a growing world population (Fageria, 2007). Rice is also the greatest consumer of water among all crops and consumes about 80% of the total irrigated fresh water resources in Asia (Bouman and Tuong, 2001;Maclean *et al.*, 2002). Fresh water, however, is becoming increasingly scarce because of the global weather changes, population growth, increasing urban and industrial development, and the decreasing availability resulting from pollution and resource depletion (Belder *et al.*, 2005; Bouman,

2007). The challenge to produce more food under increasing water scarcity has led to the notion that crop water productivity (economic yield over the amount of water consumed) needs to increase (Kijne *et al.*, 2002, 2003). However, how to increase water productivity is still being debated (Kijne *et al.*, 2003; Bouman, 2007). In cereals and at the crop level it is proposed that water productivity can be defined as the ratio of grain yield over amount of water transpired (WP_T) (Bouman, 2007). As the grain yield is the product of harvest index (*HI*) and total above-ground biomass, the WP_T in rice could be expressed as

$$WP_T = Y/T = HI \times B/T$$

where WP_T is the grain yield per unit water transpired (kg grain kg^{-1}water), *Y* is the grain yield (kg), *T* is the amount of water transpired (kg),*HI* is the harvest index (kg kg^{-1}), and *B* is the above-ground biomass (kg).

The fraction *B/T* is sometimes known as transpiration efficiency. The *HI* is the grain yield over total above-ground biomass. The grain yield and water productivity would be improved by either an increase in transpiration efficiency or an increase in *HI*. However, the ratio of biomass production over transpiration (*B/T*) has been shown to be fairly constant for a given species in a given climate (Ehlers and Goss, 2003), and can be selected for in plant breeding (Bouman, 2007). Plant biomass production is linearly coupled with the amount of water transpired, and a higher water use efficiency (*WUE*) is often a trade-off against lower biomass production (Zhang and Yang, 2004). In agriculture, many ways of conserving water have been investigated and techniques such as alternate partial rootzone irrigation, deficit irrigation, and drip irrigation, have shown that *WUE* can be enhanced (Graterol *et al.*, 1993; Zhang *et al.*, 1998; Kang *et al.*, 2000;Tabbal *et al.*, 2002; Li *et al.*, 2010). In general, these techniques are a trade-off: a lower yield for a higher *WUE* (Zhang and Yang, 2004).

On the other hand, *HI* has been shown to be a variable factor in crop production (Table 1). Variations in harvest index within a crop are mainly attributed to differences in crop management (Yang *et al.*, 2000; Guo *et al.*, 2004; Kemanian *et al.*, 2007; D'Andrea *et al.*, 2008; Peltonen-Sainio *et al.*, 2008). A water and/or nitrogen management system that could increase growth rate during grain growth and/or enhance the remobilization of assimilates from vegetative tissues to grains during the grain-filling period usually leads to a higher *HI* within a crop (Xue *et al.*, 2006; Zhang *et al.*, 2008*b*; Bueno and Lafarge, 2009; Fletcher and Jamieson, 2009; Ju *et al.*, 2009). In many situations, *HI* is closely associated with *WUE* and grain yield in wheat (*Triticum aestivum* L.) and rice (Ehdaie and Waines, 1993; Yang *et al.*, 2000, 2001*b*, 2002*b*, 2003*a*,2007; Zhang *et al.*, 2008*a*, *c*). The question arises as to whether it is possible to achieve the dual goal of increasing food production and saving water through the manipulation of *HI* in crop production. Research in rice related to this question is the focus of this paper.

Table 1: Variations of harvest index in crop production

Crop	Harvest index (kg kg^{-1})	Source
Rice (Oryza sativa L.)	0.17 – 0.56	Bueno and Lafarge, 2009; Ju et al., 2009
Wheat (Triticum aestivum L.)	0.31 – 0.53	Yang et al., 2000; Zhang et al., 2008b
Barley (Hordeum vulgare L.)	0.30 – 0.62	Kemanian et al., 2007; Peltonen-Sainio et al., 2008
Maize (Zea mays L.)	0.25 – 0.58	Guo et al., 2004; D'Andrea et al., 2008

The dominant system of paddy rice production in Asia is transplanting or direct-seeding in a field that is kept continuously flooded with 5–10 cm water throughout the growing season (Bouman and Tuong, 2001). To reduce water use in irrigated rice, water-saving regimes have been introduced such as an aerobic rice system (Bouman *et al.*, 2005; Singh *et al.*, 2008; Lampayan *et al.*, 2010), a system of rice intensification (SRI) (Uphoff and Randriamiharisoa, 2002), alternate wetting and drying (AWD) irrigation (Bouman and Tuong, 2001; Belder *et al.*, 2004 2005, 2007;Zhang *et al.*, 2008*a*), controlled soil drying during grain filling (Yang *et al.*, 2002*b*, 2003*a*; Yang and Zhang, 2006), and non-flooded mulching cultivation (Liu *et al.*, 2005; Tao *et al.*, 2006; Xu *et al.*, 2007; Zhang *et al.*, 2008*c*, 2009*b*). In the aerobic rice system, rice is grown under non-flooded, non-puddled, and non-saturated soil conditions (Bouman, 2001). Although aerobic rice needs less water at the field level than conventional lowland rice, it could not replace lowland rice in most of the rice-growing areas due to its lower grain yield and poorer taste and eating qualities (Zhang and Yang, 2004;

Lampayan *et al.*, 2010). It is suggested that the aerobic rice system could be an option for farmers in rainfed lowlands with a limited or an erratic distribution of rainfall (Bouman *et al.*, 2005; Altin *et al.*, 2006). The SRI is characterized by a set of basic management practices including transplanting young seedlings with 2–4 phyllochrons, planting a single seeding per hill, wide space planting, daily or intermittent irrigation before panicle initiation, hand or mechanical weeding, and applying nutrients to soil preferably in an organic form (Stoop *et al.*, 2002; Uphoff and Randriamiharisoa, 2002). Uphoff and Randriamiharisoa (2002) claimed that double or triple yields over those of conventional rice cultures were attained by SRI in Madagascar. Rafaralahy (2002) reported yields over 15 t ha^{-1} or even above 20 t ha^{-1} by SRI in the highlands of Madagascar. Criticisms of these reports on SRI were made by Sheehy *et al.* (2004), Dobermann (2004), and Sinclair and Cassman (2004) for the extraordinary high yield, effectiveness of SRI practices, experimental procedures, and publications. The intermittent irrigation of SRI would reduce water use in rice, but the intensive labour requirement and the high level of skill needed are likely to limit the adoption of SRI in rice production (Horie *et al.*, 2005). There are reports that the practices of controlled soil drying during grain filling, alternate wetting and moderate soil drying, and non-flooded straw mulching cultivation could not only increase *WUE*, but also maintain or even increase grain yield (Yang *et al.*, 2002*b*, 2003*a*, 2006; Zhang *et al.*, 2008*a*, *c*, 2009*a*, *b*). It would be interesting to know if and how such practices manipulate *HI* and, consequently, increase *WUE* and grain yield. Therefore the discussions in this review cover: (i) post-anthesis controlled soil drying improves the remobilization of carbon reserves and grain filling; (ii) a moderate wetting drying regime reduces redundant vegetative growth and increases grain yield; and (iii) non-flooded wheat straw mulching cultivation maintains a high grain yield and increases *WUE*.

To investigate the performances of *HI*, grain yield, and *WUE* under these crop management systems, three experiments of post-anthesis soil drying, AWD, and non-flooded mulching cultivation were conducted at the research farm of Yangzhou University, Jiangsu Province, China (32°30′ N, 119°25′ E, 21 m altitude) during the rice growing season (May–October). Two high-yielding rice cultivars currently used in local production, Zhendao 88 (*japonica*) and Liangyoupeijiu (LYPJ, *indica* hybrid), were grown in the field. Each experiment was repeated for three or five growing seasons. The methodologies applied in the experiments were described previously (Yang *et al.*, 2001*c*, 2003*a*; Zhang *et al.*, 2008*c*, 2009*a*). Briefly, in the experiment of post-anthesis soil drying, two levels of nitrogen treatments were applied. Half the plots were top-dressed with either 15 g N m^{-2} (normal amount, NN) or 30 g N m^{-2} (high amount, HN) as urea. From 9 d post anthesis to maturity, three levels of soil water potential (ψ_{soil}) were imposed on the plants of both NN and HN treatments. The well-watered (WW) treatment kept a water depth of 1–2 cm (ψ_{soil} =0) in the field by manually applying water every day. The moderate soil drying (MD) treatment maintained ψ_{soil} at –25 kilopascal (kPa), and the severe soil drying (SD) treatment maintained ψ_{soil} at –50 kPa. The ψ_{soil} in the MD and SD treatments was monitored with tension meters buried in the 15–20 cm soil depth. Tension meter readings were recorded twice a day at 10.00 h and 16.00 h. When the reading dropped to the designed value, 0.23 cm and 0.12 cm of irrigation per plot was added manually to the MD and SD treatments, respectively (Yang *et al.*, 2003*a*). A rain shelter consisting of a steel-frame covered with plastic sheet was used in each block to protect the plot during rains. Grains that developed from spikelets of 15 panicles were sampled at 4 d intervals from anthesis to maturity, dried at 70 °C to constant weight for 72 h, and weighed. The grain-filling process was fitted by the Richards' growth equation (Richards, 1959) as described by Zhu *et al.* (1988):

$$W = A / (1 + Be^{-kt})^{1/N}$$

where *W* is the grain weight (mg), *A* is the final grain weight (mg), *t* is the time after anthesis (d), and B, k, and N are coefficients determined by regression. The active grain-filling period was defined as that when *W* was 5% (t1) to 95% (t2) of *A*. The average grain-filling rate during this period was calculated from t1 to t2.

In the AWD experiment, three irrigation regimes including alternate wetting and moderate soil drying (WMD), alternate wetting and severe soil drying (WSD), and conventional irrigation (CI), were applied from 10 d after transplanting to maturity. In the WMD regime, fields were not irrigated until ψ_{soil} reached –15 kPa at 15–20 cm depth. While in the WSD regime, water was withheld until ψ_{soil} reached –30 kPa at 15–20 cm depth. The CI regime was maintained as plots with a continuous flood with a water depth of 2–3 cm until one week before harvest, which are the recommended farming practices.

The experiment of non-flooded mulching cultivation comprised four treatments: traditional flooding (TF) as control, and plastic film mulching (PM), wheat straw mulching (SM), and no mulching (NM) as non-flooded cultivation systems. The TF treatment was continuously flooded with 2–3 cm water level over the plot until one week before rice harvest in line with traditional farming practices. Plastic film, 0.007 mm thick and 1.8 m wide, was used to cover the soil in the PM treatment. Wheat straw harvested from the same field in the wheat season was used to cover the soil in the SM treatment. In all the non-flooded treatments, plots were flooded only for 6–8 d after transplanting for the re-greening of seedlings. After that, the amount of water applied to PM, SM, and NM plots was controlled at 380–440 m^3 ha^{-1} at each stage of mid-tillering, booting, flowering, and early grain-filling, respectively, when ψ_{soil} had reached –25 kPa at 15–20 cm depth and if there was no precipitation at this time.

Post-Anthesis Controlled Soil Drying Improves Remobilization of Carbon Reserves and Grain Filling

Grain filling is the final stage of growth in cereals when fertilized ovaries develop into caryopses and depends on carbon from two resources: current assimilates and assimilates redistributed from reserve pools in vegetative tissues either pre- or post-anthesis (Kobata *et al.*, 1992;Schnyder, 1993; Samonte *et al.*, 2001). The contribution of reserved assimilates in culms and leaf sheaths of rice plants is estimated at 10–40% of the final yield, depending on the cultivar and the environmental conditions (Gebbing and Schnyder, 1999; Takai *et al.*, 2005). Remobilization of reserves to the grain is critical for grain yield if the plants are subjected to water stress or if the yield potential is largely based on the high biomass accumulation (Yoshida, 1972; Ehdaie and Waines, 1996; Asseng and van Herwaarden, 2003; Plaut *et al.*, 2004).

Remobilization and transfer of the stored assimilates in vegetative tissues to the grain in monocarpic plants such as rice and wheat require the initiation of whole plant senescence (Gan and Amasino, 1997; Noodén *et al.*, 1997). Delayed whole plant senescence (i.e. plants remain green when grains are due to ripen) results in much non-structural carbohydrate (NSC) left in the straw and leads to a low *HI*. Slow grain-filling can often be associated with the delay of whole plant senescence (Zhu *et al.*, 1997; Mi*et al.*, 2002; Gong *et al.*, 2005). In China, there are currently at least three common cases where plant whole senescence is unfavourably delayed: the over-use of nitrogen fertilizers (Buresh *et al.*, 2004; Peng *et al.*, 2006), adoption of lodging-resistant cultivars that stay 'green' for too long (Yuan, 1994, 1998; Zhu *et al.*, 1997), and the introduction of hybrid rice which is too vigorous (Yang *et al.*, 2002*a*; Yuan, 2003). Their senescence is defined as unfavourably delayed because the gain from the extended grain-filling period is less than the loss due to slow grain filling and unused assimilates left in the straw (Yang and Zhang, 2006).

Usually, water stress at grain-filling time induces early senescence and shortens the grain-filling period but increases the remobilization of assimilates from the straw to grains (Kobata and Takami, 1981; Nicolas *et al.*, 1985; Palta *et al.*, 1994; Asseng and van Herwaarden, 2003; Plaut *et al.*, 2004). Can the advantage of soil drying-induced whole plant senescence and better carbon remobilization be taken to improve grain yield in situations where unfavourably delayed senescence is a problem? It has been found that a controlled soil drying or a moderate soil drying, namely plants can rehydrate overnight (Fig. 1A, B) and photosynthesis is not severely inhibited (Fig. 1C, D), imposed at mid and late grain-filling stages (from 9 d post-anthesis until maturity) can greatly enhance assimilate remobilization from vegetative tissues to grains and also accelerate the grain-filling rate (Table 2). Such a controlled soil drying does not necessarily reduce grain yield even when plants are grown under normal nitrogen conditions. Furthermore, in cases where plant senescence is unfavourably delayed, such as by the heavy use of nitrogen, the gain from enhanced remobilization and accelerated grain-filling rate can outweigh the loss of reduced photosynthesis and a shortened grain-filling period, leading to an increased grain yield and higher *HI* and *WUE* (Table 3).

Table 2: Remobilization of pre-stored assimilates in straws and grain filling rate of rice subjected to various nitrogen and soil moisture treatments

Cultivar	Treatment	Remobilized C reserve (%)	Contribution to grain (%)	Total dry matter (g m−2)	Active grain-filling period (d)	Grain-filling rate (mg d−1 grain−1)
Zhendao 88	WW-NN	28.08 d	8.78 e	1857 a	23 b	1.02 d
	WW-HN	7.72 e	2.62 f	1867 a	27 a	0.84 e
	MD-NN	66.17 b	21.14 c	1684 b	18 d	1.31 b
	MD-HN	52.94 c	16.07 d	1828 a	21 c	1.17 c
	SD-NN	74.81 a	28.61 a	1358 d	13 f	1.65 a
	SD-HN	68.66 b	24.43 b	1506 c	16 e	1.38 b
LYPJ	WW-NN	11.89 d	3.89 e	1998 ab	26 b	0.88 d
	WW-HN	2.96 e	1.05 f	2069 a	32 a	0.69 e
	MD-NN	59.11 b	18.34 c	1914 b	20 d	1.17 b
	MD-HN	49.83 c	15.27 d	1991 ab	24 c	0.99 c
	SD-NN	66.78 a	24.55 a	1572 d	17 e	1.28 a
	SD-HN	60.96 b	21.37 b	1735 c	19 d	1.16 b
LSD0.05		5.31	2.45	72	1	0.11

Table 3: Grain yield, harvest index, and water use efficiency (*WUE*) for irrigation of rice subjected to various nitrogen and soil moisture treatments

Cultivar	Treatment	Total number of spikelets ($\times 10^3 m^{-2}$)	Filled grain (%)	Grain weight (mg grain^{-1})	Grain yield (g m^{-2})	Harvest index (kg kg^{-1})	*WUE* (kg grain m^{-3})
Zhendao 88	WW-NN	38.23 a	85.59 ab	26.1 b	854 b	0.46 c	1.12 c
	WW-HN	38.31 a	83.51 b	25.1 c	803 c	0.43 d	0.98 d
	MD-NN	38.45 a	83.90 b	26.1 b	842 b	0.50 ab	1.32 b
	MD-HN	38.28 a	86.05 a	27.2 a	896 a	0.49 b	1.31 b
	SD-NN	38.04 a	77.98 d	23.8 d	706 e	0.52 a	1.41 a
	SD-HN	38.19 a	80.48 c	24.5 c	753 d	0.50 ab	1.39 a
LYPJ	WW-NN	47.79 a	72.41 b	25.4 b	879 b	0.44 b	1.11 c
	WW-HN	47.65 a	69.12 c	24.5 c	807 cd	0.39 c	0.95 d
	MD-NN	47.54 a	76.18 a	25.9 ab	938 a	0.49 a	1.36 ab
	MD-HN	47.86 a	75.66 a	26.4 a	956 a	0.48 a	1.33 b
	SD-NN	47.44 a	68.75 c	24.1 c	786 d	0.50 a	1.40 a
	SD-HN	47.59 a	71.44 b	24.5 c	833 c	0.4 a	1.39 a
$LSD_{0.05}$		1.85	2.09	0.5	38	0.02	0.04

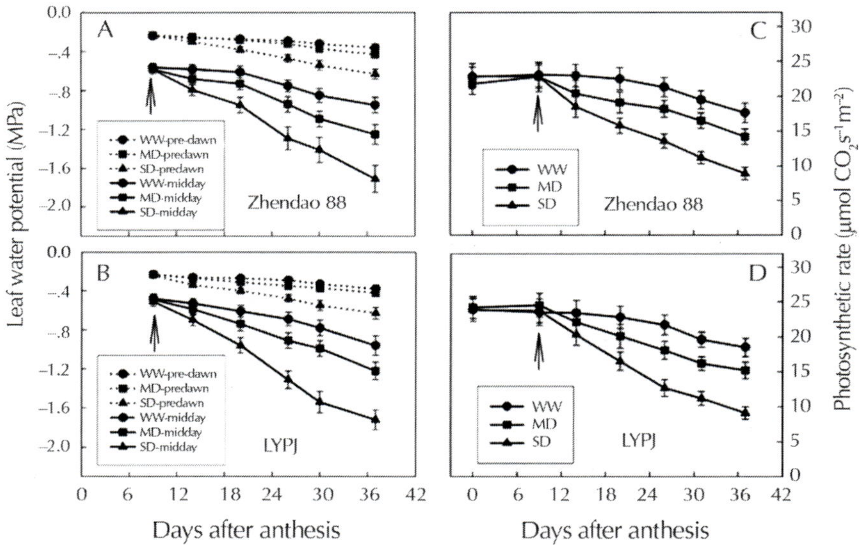

Figure 1: Leaf water potential (A, B) and photosynthetic rate (C, D) of the flag leaf of rice. The *japonica* cultivar Zhendao 88 (A, C) and *indica*hybrid Liangyoupeijiu (LYPJ) (B, D) were field grown. WW, MD, and SD are well watered [soil water potential (ψ_{soil}) = 0 kPa, moderate soil drying (ψ_{soil} = –25 kPa), and severe soil drying (ψ_{soil} = –50 kPa) treatments during grain filling. Measurements of leaf water potentials were made on the flag leaves at pre-dawn (06.00 h) and at midday (11.30 h). The photosynthetic rates were measured during 09.00–11.00 h. Arrows indicate the start of soil drying treatments. Values are averages across the three years (2005–2007). Vertical bars represent ±SE of the mean (n = 12) where these exceed the size of the symbol.

The mechanism by which post-anthesis-controlled soil drying enhances the utilization of pre-stored assimilates is not fully understood. Many processes are likely to be involved, including the hydrolysis of stored carbohydrate, phloem loading, long-distance translocation, and phloem unloading into the kernels. It was observed that, in rice stems, both α- and β-amylase activities were enhanced by the soil drying treatment, with the former enhanced more than the latter, and significantly correlated with the concentrations of soluble sugars in the stems. The other two possible starch-breaking enzymes, α-glocosidase and starch phosphorylase, showed no significant differences in the activities between well-watered and soil drying treatments. Soil drying also increased the sucrose-phosphate synthase activity that is responsible for sucrose production (Yang *et al.*, 2001*a*). In grains, the activities of four enzymes involved in sucrose-to-starch conversion: sucrose synthase, ADP glucose pyrophosphorylase (AGP), starch synthase (StS), and starch branching enzyme (SBE), were significantly enhanced by soil drying, and positively correlated with the starch accumulation rate in grains (Yang *et al.*, 2003*b*). These results suggest that, in the source (stems), enhanced activities of -amylase and sucrose-phosphate synthase contribute to the fast hydrolysis of starch and increased carbon remobilization, and in the sink side (grains), an increased grain-filling rate is mainly attributed to the enhanced sink activity by regulating key enzymes involved in sucrose-to-starch conversion, when subjected to a mild soil drying during the grain-filling period. Both abscisic acid (ABA) and cytokinins are generally believed to be two major regulators of plant senescence (Biswas and Choudhuri, 1980;Noodén, 1988; Noodén *et al.*, 1997; Haberer and Kieber, 2002). However, their regulatory roles in the remobilization of carbon reserves are not clear. The work of Yang *et al.* (2003*c*, 2004) on rice and wheat showed that the soil drying treatments substantially increased ABA accumulation (concentration) in the leaves and stems or root exudates, and markedly reduced cytokinins [zeatin (Z)+zeatin riboside (ZR)] in the leaves. Elevated ABA levels in the stems or root exudates were associated with the partitioning of pre-feed ^{14}C in the grains under soil-drying treatments. ABA in both the leaves and stems, but not cytokinins, was significantly and positively correlated with the remobilization of pre-stored carbon, and such remobilization was enhanced by exogenous ABA, suggesting that enhanced remobilization by soil drying during grain filling can be attributed, at least partly, to an elevated ABA concentration in the plant.

A MODERATE WETTING DRYING REGIME REDUCES REDUNDANT VEGETATIVE GROWTH AND INCREASES GRAIN YIELD

Sufficient water supply under the irrigated lowland rice system often leads to excessive vegetative growth which may result in less root activity, unhealthy canopy structure, and Lower *HI* (Li, 2001, Zhang and Yang, 2004). To reduce water use in irrigated rice, an alternate wetting and drying (AWD) irrigation system has been developed and is being adopted in countries of East Asia such as Bangladesh, India, Vietnam, and China (Belder *et al.*, 2004; Bouman, 2007, Zhang *et al.*, 2009*a*). In AWD, irrigation is applied a few days after water has disappeared from the surface so that periods of soil submergence alternate with periods of non-submergence during the whole growing season (Belder *et al.*, 2007). In studies on AWD irrigation, grain yield of rice was increased (Li, 2001;Tuong *et al.*, 2005; Yang *et al.*, 2007; Zhang *et al.*, 2008*a*, 2009*a*) but reduced in others (Mishra *et al.*, 1990; Tabbal *et al.*, 2002; Belder *et al.*, 2004) when compared with continuously submerged conditions. The discrepancies between the studies are probably attributed to the variations in soil hydrological conditions and the timing of the irrigation method applied (Belder *et al.*, 2004). It has been shown that an alternate wetting and moderate soil drying (WMD) regime could significantly increase both grain yield and *WUE*, and an alternate wetting and severe soil drying (WSD) regime increased *WUE*, but markedly reduced grain yield when compared with the conventional irrigation (CI) (Table 4). Increase in grain yield and*WUE* under the WMD regime could be attributed to several reasons:

Table 4: Grain yield and water use efficiency (*WUE*) for irrigation of rice under various irrigation regimes

Cultivar	Treatment	Number of panicles (m^{-2})	Number of spikelets per panicle	Filled grains (%)	Grain weight (mg $grain^{-1}$)	Grain yield (g m^{-2})	*WUE* (kg grain m^{-3})
Zhendao 88	CI	322 a	114 a	85.6 b	26.2 b	825 b	0.91 c
	WMD	319 a	116 a	90.5 a	27.2 a	908 a	1.39 a
	WSD	246 b	106 b	80.4 c	25.1 c	526 c	1.15 b
LYPJ	CI	256 a	196 a	78.6 b	25.3 b	997 b	1.01 c
	WMD	251 a	192 a	84.3 a	26.2 a	1065 a	1.51 a
	WSD	202 b	181 b	73.7 c	24.1 c	649 c	1.29 b
$LSD_{0.05}$		37	6	3.4	0.6	42	0.06

First, the WMD regime reduced redundant vegetative growth and improved canopy structure (Table 5). Compared with the CI regime, the WMD regime reduced the maximum number of tillers by 21–23%

and total leaf area by 14%, but the number of productive tillers and effective leaf area (leaf area of productive tillers) showed no significant difference between the two regimes (Table 5). As a result, the WMD regime significantly increased the percentage of productive tillers and the percentage of effective leaf area. The improved canopy quality would reduce the water used in the production of unproductive tillers and transpiration from redundant leaf area. Furthermore, the WMD regime significantly reduced the leaf angle of the top three leaves at heading time (Table 5). The improved leaf architecture would allow more radiations to penetrate the canopy, which is very important to maintain a healthy canopy during grain filling (Fageria, 2007). Although the WSD regime also reduced redundant vegetative growth and improved leaf architecture, it significantly decreased the number of productive tillers and the effective leaf area (Table 5), which may contribute to the reduction in biomass and, consequently, in grain yield.

Table 5: Maximum number of tillers, the number of productive tillers, the percentage of productive tillers, total leaf area index (*LAI*), effective *LAI*, percentage of effective *LAI*, and mean leaf angles of the top three leaves of rice under various irrigation regimes

Cultivars	Treatment	Maximum number of tillers m^{-2}	Productive tillers		Total *LAI*	Effective leaf area		Leaf angles (°)
			(Number m^{-2})	(%)		(*LAI*)	(%)	
Zhendao 88	CI	450 a	302 a	67 c	8.4 a	6.4 a	76 b	23.5 a
	WMD	345 b	297 a	86 a	7.2 b	6.3 a	88 a	20.2 b
	WSD	276 c	224 b	81 b	5.1 c	4.5 b	89 a	19.8 b
LYPJ	CI	334 a	237 a	71 c	7.8 a	5.6 a	72 b	24.2 a
	WMD	268 b	235 a	88 a	6.7 b	5.8 a	87 a	21.3 b
	WSD	224 c	184 b	82 b	4.8 c	4.1 b	86 a	20.8 b
$LSD_{0.05}$		32	16	4	0.5	0.3	2	1.1

Second, the WMD regime enhanced root and shoot activities. It was observed that the WMD regime enhanced root oxidation activity, either during the soil drying period or at the time when plants were re-watered (Fig. 2A, B), and the enhancement was more during the re-watering time. The photosynthetic rate of leaves under the WMD regime was not significantly reduced during the soil drying period, but it was significantly increased when the plant was re-watered (Fig. 2C, D). In contrast to the WMD, the WSD regime reduced the root oxidation activity and leaf photosynthetic rate, especially during the soil drying period (Fig. 2A–D). It is little understood whether the changes in photosynthetic rate are due to changes in leaf water relations, or whether there is a role for root-sourced chemical signals that alter in response to AWD. Zhang *et al.* (2009*a*)observed that changes in leaf photosynthetic rate were closely associated with changes in Z+ZR concentrations in roots under WMD and WSD regimes, implying that root-derived cytokinins may play a role in regulating leaf photosynthesis under the AWD system.

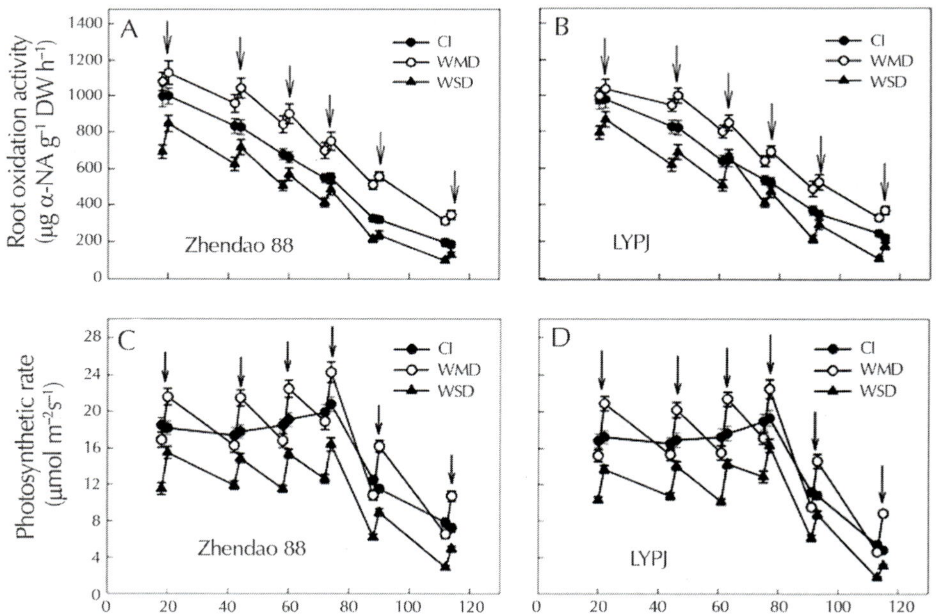

Figure 2: Root oxidation activity (A, B) and leaf photosynthetic rate (C, D) of rice. The *japonica*cultivar Zhendao 88 (A, C) and *indica* hybrid Liangyoupeijiu (LYPJ) (B, D) were field grown. CI, WMD, and WSD indicate conventional irrigation, alternate wetting and moderate soil drying, and alternate wetting and severe soil drying during the rice growing season. The photosynthetic rate was measured on the upper surface of the top fully expanded leaves at 09.00–11.00 h. Measurements of both root oxidation

activity and photosynthetic rate were made when soil water potentials were –15 kPa in WMD and –30 kPa in WSD and when plants were rewatered (indicated by arrows). Values are averages across the three years (2006–2008). Vertical bars represent ±SE of the mean (*n* = 12) where these exceed the size of the symbol.

It was also observed that under the WMD regime, the activities of three key enzymes involved in starch synthesis, AGP, StS, and SBE, in grains during the grain-filling period could be maintained or increased during soil drying, and were markedly enhanced when plants were rewatered (Table 6). The increased sink strength through enhancement in the activities of these enzymes under the WMD regime may contribute to a greater percentage of filling grains and a higher grain weight, and, consequently, to a higher grain yield. On the other hand, the reduction in grain yield under the WSD regime may be attributed to the reduced root and shoot activities and sink strength (Fig. 2; Table 6).

Table 6: Activities of adenosine diphosphate glucose pyrophosphorylase (AGP), starch synthase (StS), and starch-branching enzyme (SBE) in the grains of rice under various irrigation regimes

Cultivar	Treatment	During the soil drying period			During the rewatering time		
		AGP	StS	SBE	AGP	StS	SBE
Zhendao 88	CI	46 a	8.5 b	685 b	44 b	8.2 b	637 b
	WMD	48 a	12.6 a	767 a	57 a	16.4 a	859 a
	WSD	21 b	4.4 c	389 c	32 c	5.8 c	498 c
LYPJ	CI	42 a	9.3 b	576 b	39 b	8.8 b	554 b
	WMD	45 a	14.5 a	694 a	55 a	17.9 a	796 a
	WSD	25 b	5.7 c	321 c	30 c	7.2 b	432 c
$LSD_{0.05}$		4	1.6	48	5	2.1	57

Third, the WMD regime increased pre-stored carbon remobilization and*HI*. Biomasses showed no significant difference between CI and WMD regimes (Table 7). However, the WMD regime significantly facilitated the reallocation of pre-anthesis assimilates from straws to grains. From anthesis to maturity, the remobilized C reserves from vegetative tissues for plants under the WMD regime was as 2.5–3.0-fold that of plants under the CI regime (Table 7). The contribution of remobilized C reserves to the grain for plants under the WMD regime was increased by 10.0–12.9% when compared with that under the CI regime. The enhanced remobilization by the WMD regime led to a higher *HI* (Table 7). Although the WSD regime significantly increased pre-stored carbon remobilization and *HI*, it markedly reduced biomass production (Table 7). The loss from biomass production could not compensate the gain from the partitioning or remobilization of assimilates, leading to the reduction in grain yield.

Table 7: Remobilization of pre-stored assimilates in straws and harvest index of rice under various irrigation regimes

Cultivar	Treatment	NSC in strawsat heading (g m^{-2})	NSC in straws at maturity (g m^{2})	Remobilized C reserve (%)	Contribution to grain (%)	Total dry matter (g m^{2})	Harvest index (kg kg^{1})
Zhendao 88	CI	274 a	238 a	13.1 c	4.5 c	1713 a	0.47 b
	WMD	282 a	138 b	51.1 b	15.5 b	1820 a	0.51 a
	WSD	161 b	63 c	60.9 a	18.6 a	1012 b	0.52 a
LYPJ	CI	352 a	289 a	17.9 c	6.6 c	2127 a	0.45 b
	WMD	363 a	155 b	51.3 b	19.5 b	2130 a	0.50 a
	WSD	225 b	71 c	68.4 a	23.7 a	1273 b	0.51 a
$LSD_{0.05}$		26	17	5.3	2.1	110	0.02

Non-flooded straw mulching cultivation maintains a high grain yield and increases WUE

Non-flooded mulching cultivation, either non-flooded plastic film mulching cultivation (PM) or non-flooded wheat/rice straw mulching cultivation (SM), has been adopted and developed as a new rice production technique in recent years (Fan *et al.*, 2005; Liu *et al.*, 2005; Lu *et al.*, 2007). Both PM and SM are employed under non-flooded conditions with limited irrigation, and they are substantially different from both traditional flooded rice cultivation and rain-fed rice cultivation (Li *et al.*, 2006). There are reports that the PM, characterized by its striking efficiency in the maintenance of soil moisture, the increase in soil temperature in the early season, and the inhibition of

5. Belder P, Bouman BAM, Spiertz JHJ. Exploring options for water saving in lowland rice using a modeling approach. Agricultural. Systems 2007;92:91-114.
6. Belder P, Spiertz JHJ, Bouman BAM, Lu G, Tuong TP. Nitrogen economy and water productivity of lowland rice under water-saving irrigation. Field Crops Research2005;93:169-185.
7. Biswas AK, Choudhuri MA. Mechanism of monocarpic senescence in rice. Plant Physiology 1980;65:340-345.
8. Bouman BAM. Water-efficient management strategies in rice production.International Rice Research Notes 2001;16:17-22.
9. Bouman BAM. A conceptual framework for the improvement of crop water productivity at different spatial scales. Agricultural Systems 2007;93:43-60.
10. Bouman BAM, Peng S, Castaňeda AR, Visperas RM. Yield and water use of irrigated tropical aerobic rice systems. Agricultural Water Management 2005;74:87-105.
11. Bouman BAM, Tuong TP. Field water management to save water and increase its productivity in irrigated lowland rice. Agricultural Water Management 2001;49:11-30.
12. Bueno CS, Lafarge T. Higher crop performance of rice hybrids than of elite inbreds in the tropics. 1. Hybrids accumulate more biomass during each phenological phase. Field Crops Research 2009;112:229-237.
13. Buresh R, Peng S, Huang J, Yang J, Wang G, Zhong X, Zou Y. Rice systems in China with high nitrogen inputs. In: Mosier AR, Syers JK, Freney JR, editors. Agriculture and the nitrogen cycle: assessing impacts of fertilizer use on food production and the environment. London, UK: Island Press; 2004. p. 143-153.
14. D›Andrea KE, Otegui ME, de la Vega AJ. Multi-attribute responses of maize inbred lines across managed environments. Euphytica 2008;162:381-394.
15. Dobermann A. A critical assessment of the system of rice intensification (SRI).Agricultural Systems 2004;79:261-281.
16. Ehdaie B, Waines JG. Variation in water-use efficiency and its components in wheat. I. Well-watered pot experiment. Crop Science 1993;33:294-299.

17. Ehdaie B, Waines JG. Genetic variation for contribution of preanthesis assimilates to grain yield in spring wheat. Journal of Genetics and Breeding 1996;50:47-55.

18. Ehlers W, Goss M. Water dynamics in plant production. Wallingford, UK: CABI Publishing; 2003. p. 297-305.

19. Fageria NK. Plant tissue test for determination of optimum concentration and uptake of nitrogen at different growth stages in low land rice. Communication in Soil Science and Plant Analysis 2003;34:259-270.

20. Fageria NK. Yield physiology of rice. Journal of Plant Nutrition 2007;30:843-879.

21. Fan MS, Liu XJ, Jiang RF, Zhang FS, Lu SH, Zeng XZ, Christie P. Crop yields, internal nutrient efficiency, and changes in soil properties in rice-wheat rotations under non-flooded mulching cultivation. Plant and Soil 2005;277:265-276.

22. Fletcher AL, Jamieson PD. Causes of variation in the rate of increase of wheat harvest index. Field Crops Research 2009;113:268-273.

23. Funaba M, Ishibashi Y, Molla AH, Iwanami K, Iwaya-Inoue M. Influence of low/high temperature on water status in developing and maturing rice grains. Plant Production Science 2006;9:347-354.

24. Gan S, Amasino RM. Making sense of senescence. Plant Physiology1997;113:313-319.

25. Gebbing T, Schnyder H. Pre-anthesis reserve utilization for protein and carbohydrate synthesis in grains of wheat. Plant Physiology 1999;121:871-878.

26. Gong Y, Zhang J, Gao J, Lu J, Wang J. Slow export of photoassimilate from stay-green leaves during late grain filling stage in hybrid winter wheat (*Triticum aestivum*L.). Journal of Agronomy and Crop Science 2005;191:292-299.

27. Graterol YE, Eisenhauer DE, Elmore RW. Alternate-furrow irrigation for soybean production. Agricultural Water Management 1993;24:133-145.

28. Guo QF, Wang QC, Wang LM. Maize production in China. Shanghai: Shanghai Science & Technology Press; 2004. p. 117-167.

29. Haberer G, Kieber JJ. Cytokinins. New insights into a classic phytohomone. Plant Physiology 2002;128:354-362.
30. Hasegawa T, Fujimura S, Shimono H, Iwama K, Jiteuyama Y. Rice growth and developing limited by root zone temperature. In: Morita S, editor. Proceedings of the sixth symposium of the International Society for Root Research. Nagoya, Japan:Japanese Society for Root Research (JSRR); 2001. p. 520-521.
31. Horie T, Shiraiwa T, Homma K, Katsura K, Maeda Y, Yoshida H. Can yields of lowland rice resume the increases that showed in the 1980s? Plant Production Science 2005;8:259-274.
32. Ju J, Yamamoto Y, Wang YL, Shan YH, Dong GC, Miyazaki A, Yoshida T. Genotypic differences in dry matter accumulation, nitrogen use efficiency and harvest index in recombinant inbred lines of rice under hydroponic culture. Plant Production Science2009;12:208-216.
33. Kang S, Shi W, Zhang J. An improved water-use efficiency for maize grown under regulated deficit irrigation. Field Crops Research 2000;67:207-214.
34. Kemanian AR, Stockle CO, Huggins DR, Viega LM. A simple method to estimate harvest index in grain crops. Field Crops Research 2007;103:208-216.
35. Kijne JW, Barker R, Molden D. Water productivity in agriculture: limits and opportunities for improvement. Wallingford, UK: CABI Publishing; 2002.
36. Kijne JW, Tuong TP, Bennett J, Bouman BAM, Oweis T. Background papers for the Challenge Program for food and water. Colonbo, Sri Lanka: CGIAR-IWMI; 2003.Ensuring food security via crop water productivity improvement; p. 1-42.
37. Kobata T, Palta JA, Turner NC. Rate of development of post-anthesis water deficits and grain filling of spring wheat. Crop Science 1992;32:1238-1242.
38. Kobata T, Takami S. Maintenance of the grain growth in rice subject to water stress during the early grain filling. Japanese Journal of Crop Science 1981;50:536-545.
39. Lampayan RM, Bouman BAM, de Dios JL, Espiritu AJ, Soriana JB, Lactaoen AT,Faronilo JE, Thant KM. Yield of aerobic rice in rainfed

lowlands of the Philippines as affected by nitrogen management and row spacing. Field Crops Research2010;116:165-174.

40. Li FS, Yu JM, Nong ML, Kang SZ, Zhang JH. Partial root-zone irrigation enhanced soil enzyme activities and water use of maize under different ratios of inorganic to organic nitrogen fertilizers. Agricultural Water Management 2010;97:231-239.
41. Li YH. Research and practice of water-saving irrigation for rice in China. In:Barker R, Loeve R, Li Y, Toung TP, editors. Proceedings of an international workshop on water-saving irrigation for rice. Wuhan, China: Hubei Science Press; 2001. p.135-144.
42. Li YS, Wu LH, Lu XH, Zhao LM, Fan QL, Zhang FS. Soil microbial biomass as affected by non-flooded plastic mulching cultivation in rice. Biology and Fertility of Soils 2006;42:107-111.
43. Liu L, Yuan L, Wang Z, Xu G, Cheng Y. Preliminary studies on the physiological reason and countermeasure of lodging in non-flooded rice. Chinese Journal of Rice Science 2002;16:225-230. (in Chinese with English abstract).
44. Liu XJ, Ai YW, Zhang FS, Lu SH, Zeng XZ, Fan MS. Crop production, nitrogen recovery and water use efficiency in rice–wheat rotations as affected by non-flooded mulching cultivation (NFMC). Nutrient Cycling Agroecosystem 2005;71:289-299.
45. Liu XJ, Wang JC, Lu SH, Zhang FS, Zeng XZ, Ai YW, Peng S, Christie P. Effects of non-flooded mulching cultivation on crop yield, nutrient uptake and nutrient balance in rice–wheat cropping systems. Field Crops Research 2003;83:297-311.
46. Lu X, Wu L, Pang L, Li Y, Wu J, Shi C, Zhang F. Effects of plastic film mulching cultivation under non-flooded condition on rice quality. Journal of the Science of Food and Agriculture 2007;87:334-339.
47. Maclean JL, Dawe DC, Hardy B, Hettel GP. Rice almanac. Los Baños, Philippines:International Rice Research Institute; 2002. p. 1-10.
48. Mi G, Tang L, Zhang F, Zhang J. Carbohydrate storage and utilization during grain filling as regulated by nitrogen application in two wheat cultivars. Journal of Plant Nutrition 2002;25:213-229.
49. Mishra HS, Rathore TR, Pant RC. Effect of intermittent irrigation on groundwater table contribution, irrigation requirement and

yield of rice in Mollisols of Tarai region. Agricultural Water Management 1990;18:231-241.

50. Nicolas ME, Lambers H, Simpson RJ, Dalling MJ. Effect of drought on metabolism and partitioning of carbon in two wheat varieties differing in drought-tolerance.Annals of Botany 1985;55:727-747.
51. Noodén LD. Abscisic acid, auxin, and other regulators of senescence. In: NoodénLD, Leopold AC, editors. Senescence and aging in plants. San Diego, USA: Academic Press Inc; 1988. p. 329-368.
52. Noodén LD, Guiamet JJ, John I. Senescence mechanisms. Physiologia Plantarum1997;101:746-753.
53. Palta JA, Kobata T, Turner NC, Fillery IR. Remobilization of carbon and nitrogen in wheat as influenced by post-anthesis water deficits. Crop Science 1994;34:118-124.
54. Peltonen-Sainio P, Muurinen S, Rajala A, Jauhiainen L. Variation in harvest index of modern spring barley, oat and wheat cultivars adapted to northern growing conditions. Journal of Agricultural Science 2008;146:35-47.
55. Peng S, Buresh R, Huang J, Yang J, Zou Y, Zhong X, Wang G, Zhang F. Strategies for overcoming low agronomic nitrogen use efficiency in irrigated rice systems in China. Field Crops Research 2006;96:37-47.
56. Plaut Z, Butow BJ, Blumenthal CS, Wrigley CW. Transport of dry matte into developing wheat kernels and its contribution to grain yield under post-anthesis water deficit and elevated temperature. Field Crops Research 2004;86:185-198.
57. Ponnamperuma FN. Organic matter and rice. Los Baños, Philippines: IRRI; 1984.Straw as a source of nutrients for wetland rice; p. 117-136.
58. Qin J, Hu F, Zhang B, Wei Z, Li H. Role of straw mulching in non-continuously flooded rice cultivation. Agricultural Water Management 2006;83:252-260.
59. Raison RJ. Modification of the soil environment by vegetation fires, with particular reference to nitrogen transformations: a review. Plant and Soil1979;51:73-108.
60. Rasmussen PE, Collins HP. Long-term impacts of tillage, fertilizer

and crop residue on soil organic matter in temperate semi-arid regions. Advances in Agronomy 1991;45:93-134.

61. Richards FJ. A flexible growth function for empirical use. Journal of Experimental Botany 1959;10:290-300.

62. Rafaralahy S. An NGO perspective on SRI and its origins in Madagascar. In: UphoffN, editor. *Assessment of the system for rice intensification* (SRI). Proceedings of International Conference, Sanya, China, April 1–3, 2002. Ithaca, New York: Cornell International Institute of Food Agriculture and Development; 2002. p. 17-22.

63. Samonte SOPB, Wilson LT, McClung AM, Tarpley L. Seasonal dynamics of non-structural carbohydrate partitioning in 15 diverse rice genotypes. Crop Science2001;41:902-909.

64. Schnyder H. The role of carbohydrate storage and redistribution in the source–sink relations of wheat and barley during grain filling: a review. New Phytologist1993;123:233-245.

65. Sheehy JE, Peng S, Dobermann A, Mitchell PL, Ferrer A, Yang J, Zou Y, Zhong X,Huang J. Fantastic yields in the systems of rice intensification: fact or fallacy? Field Crops Research 2004;88:1-8.

66. Sinclair TR, Cassman KG. Agronomic UFOs. Field Crops Research 2004;88:9-10.

67. Singh S, Ladha JK, Gupta RK, Bhushan L, Rao AN. Weed management in aerobic rice systems under varying establishment methods. Crop Production 2008;27:660-671.

68. Stoop WA, Uphoff N, Kassam A. A review of agricultural research raised by the system of rice intensification (SRI) from Madagascar: opportunities for improving farming systems for resource-poor farmers. Agricultural Systems 2002;71:249-274.

69. Tabbal DF, Bouman BAM, Bhuiyan SI, Sibayan EB, Sattar MA. On-farm strategies for reducing water input in irrigated rice: case studies in the Philippines. Agricultural Water Management 2002;56:93-112.

70. Takai T, Fukuta Y, Shirawa T, Horie T. Time-related mapping of quantitative trait loci controlling grain-filling in rice. Journal of Experimental Botany 2005;56:2107-2118.

71. Tao H, Brueck H, Dittert K, Kreye C, Lin S, Sattelmacher B. Growth

and yield formation for rice (*Oryza sativa* L.) in the water-saving ground cover rice production system (GCRPS). Field Crops Research 2006;95:1-12.

72. Tuong TP, Bouman BAM, Mortimer M. More rice, less water: integrated approaches for increasing water productivity in irrigated rice-based systems in Asia.Plant Production Science 2005;8:231-241.
73. Uphoff N, Randriamiharisoa. Reducing water use in irrigated rice production with the Madagascar system of rice intensification (SRI). In: Bouman BAM, editor. Water-wise rice production. Los Baños, Philippines: International Rice Research Institute;2002. p. 151-166.
74. Xu GW, Zhang ZC, Zhang JH, Yang JC. Much improved water use efficiency of rice under non-flooded mulching cultivation. Journal of Integrative Plant Biology2007;49:1527-1534.
75. Xue QW, Zhu ZX, Musick JT, Stewart BA, Dusek DA. Physiological mechanisms contributing to the increased water-use efficiency in winter wheat under deficit irrigation. Journal of Plant Physiology 2006;163:154-164.
76. Yang J, Liu K, Wang Z, Du Y, Zhang J. Water-saving and high-yielding irrigation for lowland rice by controlling limiting values of soil water potential. Journal of Integrative Plant Biology 2007;49:1445-1454.
77. Yang J, Peng S, Zhang Z, Wang Z, Visperas RM, Zhu Q. Grain and dry matter yields and partitioning of assimilates in *japonica/indica* hybrids. Crop Science2002a;42:766-772.
78. Yang J, Zhang J. Grain filling of cereals under soil drying. New Phytologist2006;169:223-236.
79. Yang J, Zhang J, Huang Z, Zhu Q, Wang L. Remobilization of carbon reserves is improved by controlled soil-drying during grain filling of wheat. Crop Science2000;40:1645-1655.
80. Yang J, Zhang J, Liu L, Wang Z, Zhu Q. Carbon remobilization and grain filling in*japonica/indica* hybrid rice subjected to post-anthesis water deficits. Agronomy Journal 2002b;94:102-109.
81. Yang J, Zhang J, Wang Z, Liu L, Zhu Q. Post-anthesis water deficits enhance grain filling in two-line hybrid rice. Crop Science 2003a;43:2099-2108.

82. Yang J, Zhang J, Wang Z, Zhu Q. Activities of starch hydrolytic enzymes and sucrose-phosphate synthase in the stems of rice subjected to water stress during grain filling. Journal of Experimental Botany 2001a;52:2169-2179.
83. Yang J, Zhang J, Wang Z, Zhu Q, Liu L. Water deficit-induced senescence and its relationship to the remobilization of pre-stored carbon in wheat during grain filling.Agronomy Journal 2001b;93:196-206.
84. Yang J, Zhang J, Wang Z, Zhu Q, Liu L. Activities of enzymes involved in sucrose-to-starch metabolism in rice grains subjected to water stress during grain filling.Field Crops Research 2003b;81:69-81.
85. Yang J, Zhang J, Wang Z, Zhu Q, Liu L. Involvement of abscisic acid and cytokinins in the senescence and remobilization of carbon reserves in wheat subjected to water stress during grain filling. Plant, Cell and Environment 2003c;26:1621-1631.
86. Yang J, Zhang J, Wang Z, Zhu Q, Wang W. Remobilization of carbon reserves in response to water deficit during grain filling of rice. Field Crops Research2001c;71:47-55.
87. Yang J, Zhang j, Ye Y, Wang Z, Zhu Q, Liu L. Involvement of abscisic acid and ethylene in the responses of rice grains to water stress during filling. Plant Cell and Environment 2004;27:1055-1064.
88. Yoshida S. Physiological aspects of grain yield. Annual Review of Plant Physiology1972;23:437-464.
89. Yuan LP. Increasing yield potential in rice by exploitation of heterosis. In: VirmaniSS, editor. Hybrid rice technology, new developments and future prospects. Los Baños, Philippines: International Rice Research Institute; 1994. p. 1-6.
90. Yuan LP. Hybrid rice breeding in China. In: Virmani SS, Siddiq EA, Muralidharan K, editors. Advances in hybrid rice technology. *Proceedings of the Third International Symposium on hybrid rice, Hyderabad, India, 14–16 November 1996*. Los Baños, Philippines: International Rice Research Institute; 1998. p. 27-33.
91. Yuan LP. Recent progress in breeding super hybrid rice in China.

In: Virmani SS,Mao CX, Harby B, editors. Hybrid rice for food security, poverty alleviation, and environmental protection. Los Baños, Philippines: International Rice Research Institute; 2003. p. 3-6.

92. Zhang H, Xue Y, Wang Z, Yang J, Zhang J. An alternate wetting and moderate soil drying regime improves root and shoot growth in rice. Crop Science2009a;49:2246-2260.
93. Zhang H, Zhang S, Zhang J, Yang J, Wang Z. Post-anthesis moderate wetting drying improves both quality and quantity of rice yield. Agronomy Journal2008a;100:726-734.
94. Zhang J, Sui X, Li B, Su B, Li J, Zhou D. An improved water-use efficiency for winter wheat grown under reduced irrigation. Field Crops Research 1998;59:91-98.
95. Zhang J, Yang J. Crop yield and water use efficiency. In: Bacon MA, editor. Water use efficiency in plant biology. Oxford, UK: Blackwell Publishing; 2004. p. 189-218.
96. Zhang XY, Chen SY, Sun HY, Pei D, Wang YM. Dry matter, harvest index, grain yield and water use efficiency as affected by water supply in winter wheat. Irrigation Science 2008b;27:1-10.
97. Zhang Z, Xue Y, Wang Z, Yang J, Zhang J. The relationship of grain filling with abscisic acid and ethylene under non-flooded mulching cultivation. Journal of Agricultural Science 2009b;147:423-436.
98. Zhang Z, Zhang S, Yang J, Zhang J. Yield, grain quality and water use efficiency of rice under non-flooded mulching cultivation. Field Crops Research 2008 c;108:71-81.
99. Zhu Q, Cao X, Luo Y. Growth analysis in the process of grain filling in rice. Acta Agronomica Sinica 1988;14:182-192.
100. Zhu Q, Zhang Z, Yang J, Wang Z. Source–sink characteristics related with the yield of inter-subspecific hybrid rice. Scentia Agricultura Sinica 1997;30:52-59.

Chapter 7

Nitrogen Management is Essential to Prevent Tropical Oil Palm Plantations from Causing Ground-level Ozone Pollution

C. N. Hewitta,[1], A. R. MacKenzie[a], P. Di Carlo[b], C. F. Di Marco[c], J. R. Dorsey[d], M. Evans[e], D. Fowler[c], M. W. Gallagher[d], J. R. Hopkins[f], C. E. Jones[f], B. Langford[a], J. D. Lee[f], A. C. Lewis[f], S. F. Lim[g],[2], J. McQuaid[e], P. Misztal[c], S. J. Moller[f], P. S. Monks[h], E. Nemitz[c], D. E. Oram[i], S. M. Owen[c], G. J. Phillips[c], T. A. M. Pugh[a], J. A. Pyle[j], C. E. Reeves[i], J. Ryderc, J. Siong[c], U. Skiba[c], and D. J. Stewart[i],[3]

[a]Lancaster Environment Centre, Lancaster University, Lancaster, LA1 4YQ, United Kingdom;

[b]Dipartimento di Fisica, University of L'Aquila, 67010 L'Aquila, Italy;

[c]Centre for Ecology and Hydrology, Penicuik, Edinburgh EH26 0QB, United Kingdom;

[d]School of Earth, Atmospheric and Environmental Sciences, University of Manchester Manchester, M13 9PL, United Kingdom;

[e]School of Earth and Environment, University of Leeds, Leeds LS2 9JT, United Kingdom;

[f]Department of Chemistry, University of York, York YO10 5DD, United Kingdom;

[g]Malaysian Meteorological Department, Jalan Sultan, Petaling Jaya, Selangor Darul Ehsan, Malaysia;

[h]Department of Chemistry, University of Leicester, Leicester LE1 7RH, United Kingdom;

[i]School of Environmental Sciences, University of East Anglia, Norwich NR4 7TJ, United Kingdom; and

[j]Department of Chemistry, University of Cambridge, Cambridge CB2 1EW, United Kingdom

ABSTRACT

More than half the world's rainforest has been lost to agriculture since the Industrial Revolution. Among the most widespread tropical crops is oil palm (Elaeis guineensis): global production now exceeds 35 million tonnes per year. In Malaysia, for example, 13% of land area is now oil palm plantation, compared with 1% in 1974. There are enormous pressures to increase palm oil production for food, domestic products, and, especially, biofuels. Greater use of palm oil for biofuel production is predicated on the assumption that palm oil is an "environmentally friendly" fuel feedstock. Here we show, using measurements and models, that oil palm plantations in Malaysia directly emit more oxides of nitrogen and volatile organic compounds than rainforest. These compounds lead to the production of ground-level ozone (O_3), an air pollutant that damages human health, plants, and materials, reduces crop productivity, and has effects on the Earth's climate. Our measurements show that, at present, O_3 concentrations do not differ significantly over rainforest and adjacent oil palm plantation landscapes. However, our model calculations predict that if concentrations of oxides of nitrogen in Borneo are allowed to reach those currently seen over rural North America and Europe, ground-level O_3 concentrations will reach 100 parts per billion (109) volume (ppbv) and exceed levels known to be harmful to human health. Our study provides an early warning of the urgent need to develop policies that manage nitrogen emissions if the detrimental effects of palm oil production on air quality and climate are to be avoided.

Ground-level ozone (O_3) is a priority air pollutant that damages human health, plants, and materials, reduces crop productivity, and has direct and indirect effects on the Earth's climate system (1). It is formed in the atmosphere by reactions involving oxides of nitrogen (NO_x) and volatile organic compounds (VOCs) in the presence of sunlight. The terrestrial biosphere is a major source of both these families of trace gases; in fact, the great majority of reactive VOCs globally are of biogenic origin (2). Here we show, using integrated and fully comprehensive measurements of biosphere-to-atmosphere trace gas fluxes and atmospheric composition, together with atmospheric chemistry modeling, that conversion of tropical rainforest to oil palm plantations results in much greater emissions of these reactive trace gases that lead to O_3 formation. Increased NO_x emissions will cause severe ground-level O_3 pollution (> 100 ppbv), but this pollution could be prevented by strict control of emissions of reactive nitrogen species to the atmosphere. Our study shows the importance of quantifying the current and future effects of land use change on air quality when assessing the "environmental friendliness" of palm oil and other biofuel crops. Of course, air quality is only a single consideration; in assessing the consequences of biofuel production, effects on greenhouse gas emissions and climate change, deforestation, biodiversity, water pollution and freshwater availability, and food prices and food security are all important (3); these factors are not considered here. Specifically, our study provides an early warning of the urgent need to develop policies that manage nitrogen emissions to the atmosphere from the tropics if the detrimental effects of palm oil production on air quality and climate are to be avoided.

RESULTS

Our study comprises a truly integrated and fully comprehensive set of biosphere-to-atmosphere flux measurements, atmospheric composition measurements, and atmospheric chemistry modeling in the tropics for 2 distinct but contiguous land use types. The project was based in Sabah, Malaysian Borneo, during April through July

2008. Our measurements (Table 1) show that emissions of VOCs from the rainforest and oil palm plantation landscapes are dominated by emissions of isoprene (2-methyl-1,3-butadiene). On a land area basis, isoprene emissions from the plantation (27 tonnes of isoprene per km2 per year) are 5 times greater than from the rainforest. Isoprene and other VOC emission rates depend, in part, on leaf temperature (4). Temperatures were higher at the oil palm site than at the rainforest site (campaign averages 27.9°C and 25.7°C, respectively), but this difference in temperature explains only 7% of the observed difference in isoprene emission rates; the remainder is the result of the higher biogenic VOC emission rates from oil palm (Elaeis guineensis) trees (5) compared with rainforest tree species. Isoprene has a 5 times greater potential for photochemical ozone creation than the weighted average of VOC compounds emitted by urban anthropogenic activity (6). After normalizing for this difference in VOC reactivity, the effective emissions of VOCs from the oil palm canopy per unit land area exceed those of a typical European city, such as London (7).

Table 1: Fluxes of isoprene, monoterpenes, reactive oxides of nitrogen, and nitrous oxide

Site		Isoprene: flux at the canopy scale (10.00–16.00 h) (mg/m²/h)	Monoterpenes: flux at the canopy scale (10.00–16.00 h) (mg/m²/h)	Reactive oxides of nitrogen: flux from soils (00.00–24.00 h) (mg N/m²/h)	Reactive oxides of nitrogen: flux at the landscape scale, inferred from model (00.00–24.00 h) (mg N/m²/h)	Nitrous oxide: flux from soils (00.00–24.00 h) (mg N_2O–N/m²/h)
Rainforest	Mean	1.55 ± 0.39	0.38 ± 0.20	0.14 ± 0.115	0.009	0.036 ± 0.041
	Variability	1.22	0.34	0.035		N/A
	n	619	619	5509		89
Oil palm plantation	Mean	7.77 ± 1.94	0.09 ± 0.20	N/A	0.019	0.051 ± 0.040
	Variability	7.68	0.35			N/A
	n	164	37			101

Our observations show that the plantation landscape and associated agro-industrial activities give rise to NO_x emissions several (≈2.5) times greater than from the rainforest on a land area basis (Table 1).

The 3 major NO_x sources responsible for this increase are all linked to agro-industrial activity—vehicle exhaust, combustion at the palm oil processing plant, and substantial soil nitrogen fertilization in the plantations (≈0.5 tonnes of nitrogen per hectare per year). Aircraft measurements of acetylene, a well-established combustion tracer (8), show no significant enhancement in the boundary layer over the plantation landscape compared with the rainforest landscape. This finding is a strong indication that non-combustion sources make the major contribution to the elevated NO_x concentrations observed over the plantation landscape. Further corroboration of enhanced denitrification of oil palm plantation soils is given by the N_2O emissions from these soils, which are ≈50% larger than those from the rainforest soils (Table 1).

As well as dominating VOC emission flux rates, isoprene is the dominant (> 80%) reactive VOC by atmospheric mixing ratio in both landscapes. The concentrations of isoprene were 2 to 5 times higher over the plantation landscape than over the rainforest (Fig. 1), consistent with the observations of fluxes. The C_{10} monoterpenes are the second most important group of biogenic VOCs. At both the rainforest and the plantation sites, γ-terpinene, α-pinene, and limonene dominate the emissions and concentrations of the C_{10} compounds, giving an approximately equal split between more and less reactive monoterpenes (9).

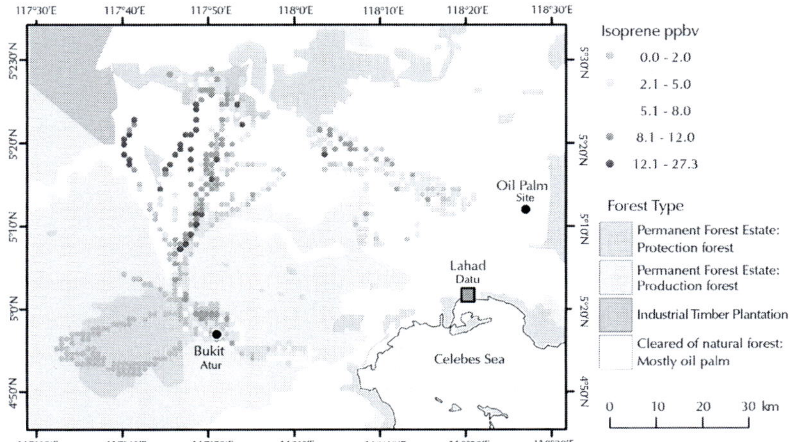

Figure 1: The aircraft flight tracks in the boundary layer for the data averaged in Fig. 2, colored by isoprene concentrations (ppbv). The markers are at

10-s intervals with the 15-s isoprene data merged onto this time stamp. Each marker represents a single measurement, and there are only a small number of instances, mostly around 5.05° North, 117.8° East, where several points overlay each other.

Fig. 2 shows that the differences in emission rates in NO_x and isoprene between the rainforest and the oil palm plantations give rise to differences in concentrations throughout the boundary layer and into the free troposphere. Over the rainforest, median boundary layer mixing ratios of NO, NO_2, and isoprene were 35, 161, and 767 parts per trillion (10^{12}) by volume (pptv), respectively, and over the plantation the median mixing ratios were 67, 288, and 3870 pptv, respectively. Surprisingly, ozone concentrations were similar over both landscapes in the boundary layer (rainforest: 11 ppbv; oil palm plantation: 12 ppbv). The ozone measurements at the Bukit Atur Global Atmosphere Watch station confirm that our measurement period is typical of the long-term record at this site. However, the concentrations of peroxyacetyl nitrate (PAN) increased from 6 pptv in the boundary layer over the rainforest to 12 pptv over the plantation landscape, demonstrating enhanced photochemical processing of the increased NO_x and VOC emissions and concentrations over the plantations.

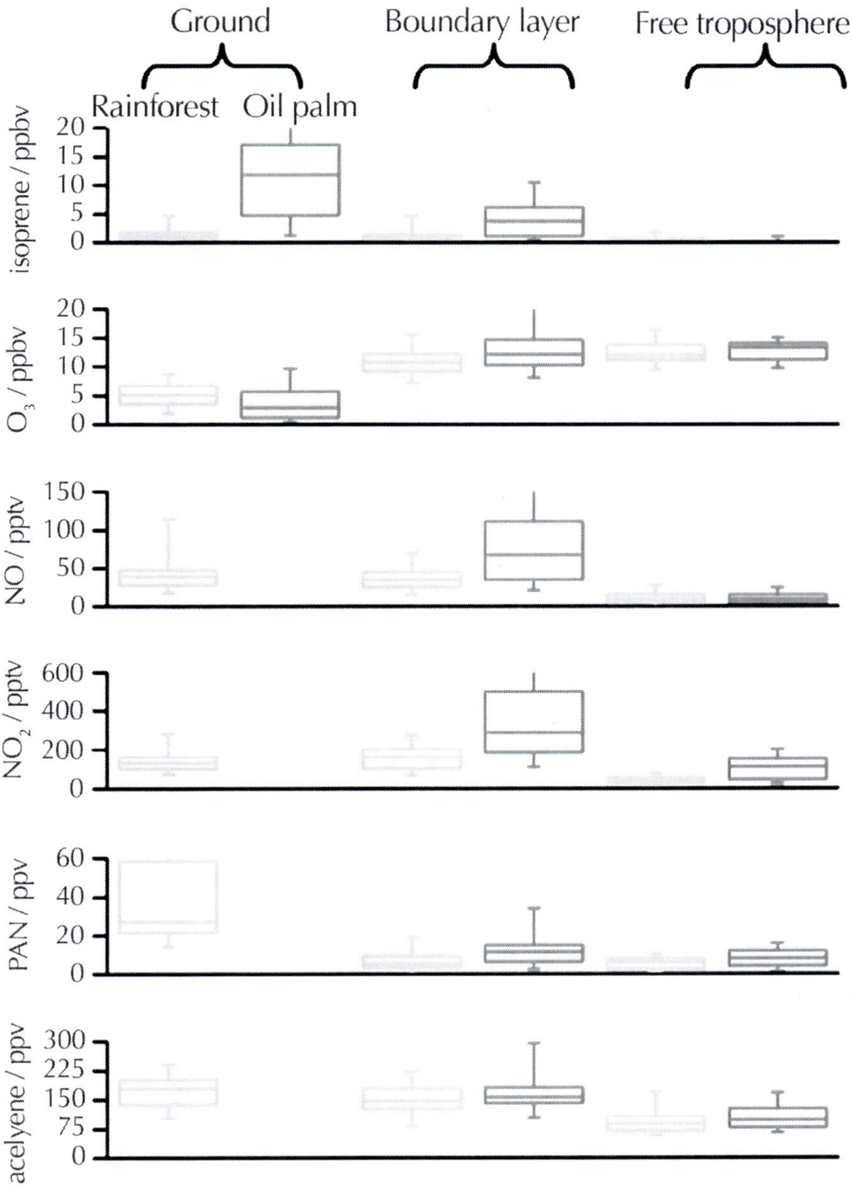

Figure 2: Isoprene, ozone, nitric oxide, nitrogen dioxide, peroxyacetyl nitrate, and acetylene measured between 10.00 and 16.00 hours local time at ground sites (2 left columns), from the aircraft flying within the boundary layer (500–800 m above ground level) (2 middle columns), and from the aircraft flying in the free troposphere (3,000–6,000 m above ground level) (2 right columns). The green symbols show data from the rainforest landscape, and the red symbols show data from the oil palm plantation landscape.

Symbol boxes show median and quartile values, with the whiskers showing 95th and 5th percentiles. Measurements of NO, NO_2, PAN, and acetylene were not made on the ground at the Sabahmas oil palm plantation and therefore do not appear in the figure. Information on sampling frequency and averaging is given in the SI.

To understand current and future ozone levels over these 2 landscapes, we have used the CiTTyCAT box model of atmospheric photochemistry (10–12). These computations explore the sensitivity of atmospheric composition in the boundary layer to VOC and NO_x emission rates for the calm, nonprecipitating, atmosphere, consistent with conditions during the measurement period.

Fig. 3 shows the model-calculated average boundary-layer ozone concentrations between 10:00 and 16:00 hours as a function of isoprene, monoterpene, and NO_x emission rates. The model has been enhanced by the inclusion of monoterpene oxidation to capture the rainforest chemistry. Monoterpenes are not yet generally included in atmospheric chemistry models such as those used in the Fourth Assessment Report of the Intergovernmental Panel on Climate Change (13).

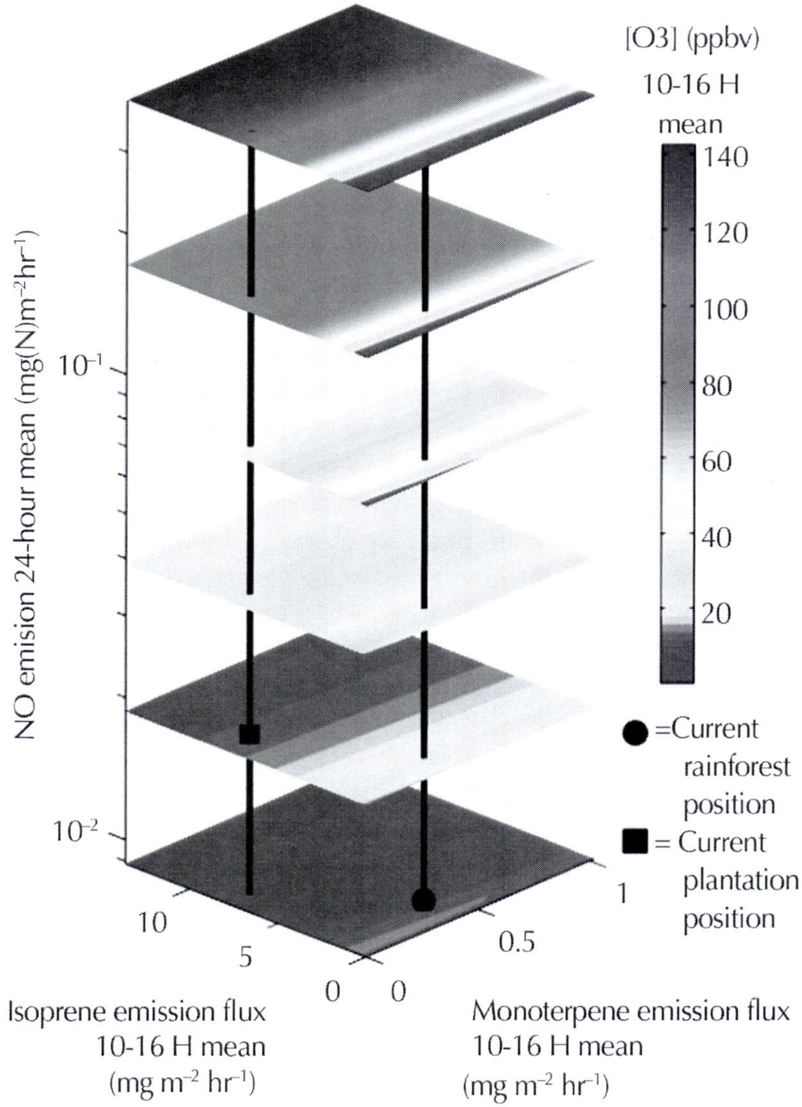

Figure 3: Sensitivity of mean mixed boundary layer ozone concentrations (from 10.00–16.00 hours) to fluxes of isoprene, monoterpenes, and NO, as computed by CiTTyCAT. VOC fluxes are the mean over 10.00–16.00 hours; NO fluxes are the mean over a 24-hour period. The black lines show the rainforest and plantation VOC emission scenarios. The current rainforest and plantation positions are determined by observed VOC emission rates and ozone concentrations.

Although the 2 landscapes currently have different NO_x and VOC concentrations, the model predicts similar ozone concentrations, as confirmed by our measurements. The ozone concentrations are similar because ozone production, although driven by increased NO and VOC emissions and oxidation, is moderated here by the rapid sequestering of reactive nitrogen to form organic nitrates that deposit to the surface and by ozone destruction resulting from its direct reaction with monoterpenes.

Fig. 3 also suggests that ozone concentrations at the 2 locations respond quite differently to increasing NO_x emissions and concentrations. In both landscapes, NO_x concentrations will increase with increasing fossil fuel consumption, associated with industrialization, across the region. Additionally, NO_x concentrations will increase in the plantation landscape with increased mechanization, fossil fuel use, and fertilizer application. Total VOC emissions in these landscapes will not be so sensitive to industrialization because of the current large biogenic VOC emission rates (Table 1).

Fig. 4 shows the sensitivity of ozone to NO_x concentrations for the VOC emission rates specific to the 2 locations. The nonlinearity in the chemistry becomes more apparent when the model results are presented this way. Our model predicts that ozone concentrations in the rainforest most probably will not exceed 50 ppbv (except by advection of ozone and its precursors from outside the rainforest landscape), irrespective of NO_x concentration, whereas in the oil palm landscape ozone concentrations will exceed 100 ppbv when NO_x concentrations reach 3–5 ppbv. Such NO_x concentrations are currently widely observed in rural areas of Western Europe (14), North America, and Asia (15). Ozone at 100 ppbv has significant detrimental effects on human health and crop yields and enhances global warming through its strong radiative forcing (1).

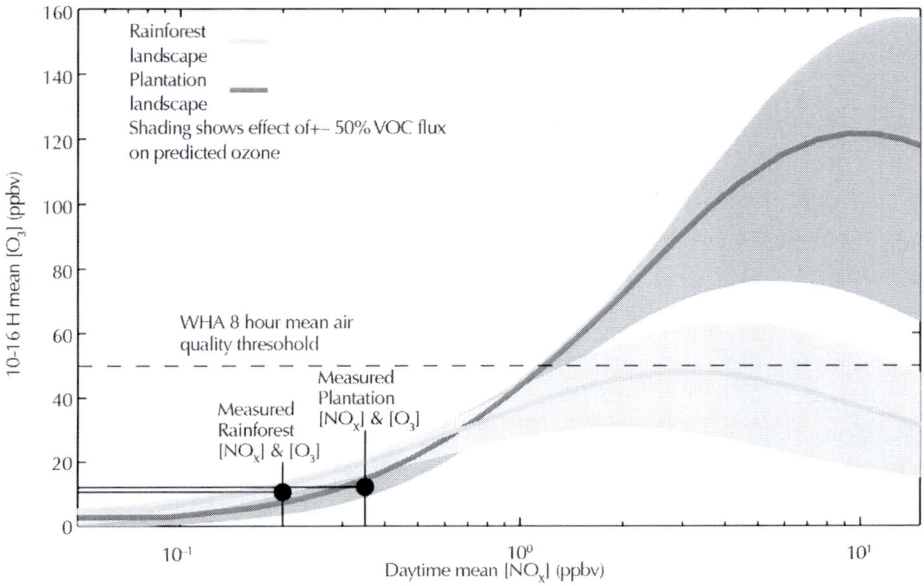

Figure 4: Sensitivity of daytime (10.00–16.00 hours) average ozone concentration to concentrations of NO_x in the boundary layer, for the isoprene and monoterpene emission rates measured at the rainforest and oil palm landscapes, as computed by CiTTyCAT. Each model run has a similar shape but with a different point of inflection, so the solid lines are not always in the centre of the shaded region. Current measured concentrations of NO_x and O_3 in the rainforest and plantation are marked.

DISCUSSION

Our measurements show that the conversion of rainforest to oil palm plantation substantially increases VOC and NO_x emissions and concentrations. The increase in VOC emission rates comes directly from the oil palm plants. The increase in NO_x emission rates comes from agro-industrialization, especially from fertilized soils. These differences in emission rate have increased the concentrations of some photochemical pollutants, notably PAN. Fortunately, they have not, to date, increased the ground-level concentrations of the most hazardous photochemical air pollutant, ozone. However, ozone will increase substantially in this region as NO_x emissions rise with industrialization and economic development.

The degree to which oil palm production will contribute to poor air quality in the future will depend almost entirely on the emission rates of nitrogen oxides. Although ozone concentrations in the rainforest are predicted to remain below 50 ppbv, regardless of the development path taken with respect to emissions of NO_x, there is the potential for ozone concentrations in the plantation landscape to greatly exceed the current World Health Organization 8-hour mean air quality threshold of 50 ppbv. Only stringent controls on emissions to maintain NO_x concentrations below those currently observed in rural parts of Western Europe and North America will prevent this increase from occurring. Controlling nitrogen emissions at both the local (plantation) and regional scales is necessary to prevent large-scale palm oil production from having very serious effects on regional air quality. Genetic modification of the oil palm cultivar to extinguish isoprene emissions (as already demonstrated for poplar [ref. 16]) would also avoid ozone air quality exceedances at any NO_x concentration (see Fig. 3). Without NO_x emissions controls or genetic modification, palm oil production for biofuel and other uses will incur significant future costs in terms of human health effects and crop yield reductions that have yet to be included in sustainability criteria for the industry. The utility of palm oil as an environmentally friendly crop may be therefore severely time-limited.

METHODS

The project was based in Sabah, Malaysia, April through July, 2008. Ground-based measurements were performed at the 100-m-high Global Atmosphere Watch tower at Bukit Atur (4° 58′ 49.33″ North, 117° 50′ 39.05″ East, 426 m above sea level) and at the Sabahmas oil palm plantation (5.2° North, 118.45° East), 70 km east-northeast of Bukit Atur. Airborne measurements were made from the Natural Environment Research Council/U.K. Meteorological Office›s BAe 146–301 large atmospheric research aircraft, which was based at Kota Kinabalu International Airport. Kota Kinabalu (population 580,000) is the largest city in Sabah and is 250 km northwest of Bukit Atur. Sandakan (population 430,000) and Tawau (population 370,000) are about 100 km north and south of Bukit Atur, respectively. The southern part of Sabah and the adjoining region of Indonesia are largely pristine, and selectively logged, rainforest and local anthropogenic emissions are insignificant. The lowland parts of northern and coastal Sabah are largely oil palm plantations. These plantations are serviced by processing mills, mainly powered by biomass (oil palm husks). No attempt was made to avoid or preferentially to sample the visible mill plumes from the aircraft. The boundary layer was sampled from the aircraft using low-level runs following the terrain, originating above the plantations and terminating above the pristine rain forest. Flying time totaled 68 hours. Measurements included trace gas concentrations and fluxes. Modeling was carried out using the CiTTyCAT box model of atmospheric chemistry. Full details of the measurement and modeling methods used are given in the supporting information (SI).

Acknowledgments

We dedicate this paper to the memory of Kate Furneaux, a core member of the OP3 project team, who was killed in a traffic accident while cycling in Leeds on 28 July 2009. We thank the Malaysian and Sabah Governments for their permission to conduct research in Malaysia; the Malaysian Meteorological Department for access to the Bukit Atur Global Atmosphere Watch station and its long-term ozone record; Wilmar International Ltd for access to and logistical support at their PPB Oil Palms Bhd Sabahmas Estate; Waidi Sinun of Yayasan Sabah and his staff, Glen Reynolds of the Royal Society's South East Asian Rainforest Research Programme and his staff, and Nick Chappell and Brian Davison of Lancaster University for logistical support at the Danum Valley Field Centre; the ground staff, engineers, scientists, and pilots of the Natural Environment Research Council/ U.K. Meteorological Office's BAe 146–301 large atmospheric research aircraft; Nick Chappell, Gemma Davies, Virginia Nicolas-Perea, and M.H. Phua for drawing and validating Fig. 1; and all the members of the OP3 project team for their support and individual and collective efforts. The work was funded by the Natural Environment Research Council Grants NE/D002117/1 and NE/E011179/1.

REFERENCES

1. The Royal Society, London (2008) Science Policy Report 15/08. Ground-level ozone in the 21st century: Future trends, impacts and policy implications (The Royal Society, London).
2. Guenther AB, et al. (1995) A global model of natural volatile organic compound emissions. J Geophys Res-Atmos 100:8873–8892.
3. Howarth RW, Bringezu S (2008) in Proceedings of the Scientific Committee on Problems of the Environment (SCOPE) International Biofuels Project Rapid Assessment (September 22–25 2008, Gummersbach, Germany), Biofuels: Environmental Consequences and Interactions with Changing Land Use (Cornell U Press, Ithaca, NY) Available at http://cip.cornell.edu/biofuels/. Accessed 07/01/09.
4. Laothawornkitkul J, Taylor JE, Paul ND, Hewitt CN (2009) Biogenic volatile organic compounds in the Earth system: A Tansley review. New Phytol 183:27–51.
5. Wilkinson MJ, et al. (2006) Circadian control of isoprene emissions from oil palm (Elaeis guineensis) Plant J 47:960–968.
6. Derwent RG, Jenkin ME, Passant NR, Pilling MJ (2007) Photochemical ozone creation potentials (POCPs) for different emission sources of organic compounds under European conditions estimated with a Master Chemical Mechanism. Atmospheric Environment 41:2570–2579.
7. Emissions of air pollutants in the UK, National Atmospheric Emissions Inventory (2009) Available at http://www.naei.org.uk/. Accessed 04/01/09.
8. Whitby RA, Altwicker ER (1978) Acetylene in the atmosphere: Sources, representative ambient concentrations and ratios to other hydrocarbons. Atmospheric Environment 12:1289–1296.
9. Atkinson R, Arey J (2003) Gas-phase tropospheric chemistry of biogenic volatile organic compounds: A review. Atmospheric Environment 37(S2):197–219.
10. Wild O, et al. (1996) Photochemical trajectory modeling studies of the North Atlantic region during August 1993. J Geophys Res 10(1):29269–29288.

11. Evans MJ, et al. (2000) Evaluation of a Lagrangian box model using field measurements from EASE (Eastern Atlantic Summer Experiment) 1996. Atmospheric Environment 34:3843–3863.
12. Donovan RG, Hope ES, Owen SM, Mackenzie AR, Hewitt CN (2005) Development and application of an urban tree air quality score for photochemical pollution episodes using the Birmingham, United Kingdom, area as a case study. Environ Sci Technol 39:6730–6738.
13. Denman KL, et al. (2007) in Climate Change 2007: The Physical Science Basis. Contribution of Working Group I to the Fourth Assessment Report of the Intergovernmental Panel on Climate Change, Couplings between changes in the climate system and biogeochemistry, ed Solomon S, et al. (Cambridge Univ Press, Cambridge, UK), pp 500–587.
14. Tarrason L, Nyiri A, eds (2008) Transboundary Acidification, Eutrophication and Ground Level Ozone in Europe in 2006 (European Monitoring and Evaluation Programme Status Report 1, Oslo).
15. Collins WJ, Stevenson DS, Johnson CE, Derwent RG (1997) Tropospheric ozone in a global-scale 3-dimensional Lagrangian model and its response to NOX emission controls. Journal of Atmospheric Chemistry 26:223–274.
16. Behkne K, et al. (2007) Transgenic, non-isoprene emitting poplars don't like it hot. Plant J 51:485–499.

Chapter 8

Effect of Irrigation Regimes and Nitrogen Levels on the Growth and Yield of Wheat

S. M. Shirazi[1], Zulkifli Yusop[1], N. H. Zardari[1], and Z. Ismail[2]

[1]Institute of Environmental and Water Resource Management (IPASA), Universiti Teknologi Malaysia (UTM), 81310 Skudai, Johor, Malaysia

[2]Department of Civil Engineering, Faculty of Engineering, University of Malaya, 50603 Kuala Lumpur, Malaysia

ABSTRACT

A field experiment was carried out to evaluate the effect of irrigation regimes and nitrogen levels on the growth and yield of wheat cv. Kanchan (*Triticum aestivum* L.). The experiment includes two factors such as four irrigation regimes and four nitrogen levels. Three farmer's fields were selected for experimentation as replication. Yield and yield contributing factors were significantly affected by irrigation regimes and different doses of nitrogen. Maximum grain yield of 2.27 t ha^{-1} by the application of 200 mm irrigation treatment. Interaction between 200 mm irrigation and 120 kg N ha^{-1} was the best combination treatment.

Introduction

Wheat is the second most important cereal crop and covers 0.64 million ha of land with an annual production of 1.2 million tons in Bangladesh [1]. Area under wheat is increasing because wheat requires less amount of irrigation compared to other cereal crops like rice. Wheat is grown in the driest months of the year when rainfall is scarce. Water balance analysis at Mymensingh region in Bangladesh depicts that during January to April and November to December potential evapotranspiration (PET) is higher than precipitation (Figure 1). Irrigation is necessary in order to grow crops during this period because of insufficient amount of rain water and high atmospheric evaporative demand by crops (Figure 2). Nitrogen (N) is a key element for plant nutrition. Applying N and phosphorus (P) fertilizers and other management practices increased the yield of wheat but in some cases these show adverse effects due to severely limiting irrigation [2, 3]. Nitrogen use efficiency can be increased by combining fertilizer, soil, water, and management. Two main approaches can be undertaken: increasing the use of N during crop growing season and decreasing the losses of N by applying optimum doses [4]. Proper growth and development of wheat needs favorable soil moisture in the root zone. Extractable water capacity of soil has significant influence on wheat grain yield and water productivity response to irrigation [5]. The moisture content in the soil gradually decreases with time in dry season and simultaneously soil moisture tension increases. Excessive irrigation increases evapotranspiration and decreases water use efficiency and may also reduce grain yield [6]. Limited irrigation is an important constraint for wheat production in rainfed, tropical, arid, and semi-arid regions. In scarce rainfall conditions and less irrigation, the best management option for N and irrigation levels is to maintain the maize-wheat cropping sequence [7] and limited irrigation water is combined with N fertilizer [8] to get maximum productivity of wheat. Supplemental irrigation significantly increased the yield of wheat with respect to rainfed treatment. According to Karam et al. [9], about 50% of soil water deficit as supplemental irrigation and 150 kg N ha^{-1} was the optimum combination for maximum grain yield of wheat. Level of fertilization and irrigation is very important to increase the fertilizer efficiency and decrease the loss of water. At a very high tension (about 15 bar), plants cannot absorb water from the soil through root zone and as a result it has an ultimate impact on crop yield. Proper timing and frequency of supplementary irrigation in relation to crop yield are crucial in irrigation scheduling for the most effective use of available water in optimizing wheat production [10–19]. Lack of irrigation at heading and grain formulation and during maturation significantly reduced the yield of wheat grain [20–26]. Grain yield increased with increase in the frequency of irrigation [27]. Improper scheduling of irrigation results not only in wastage of water but also in decreasing crop growth and yield [28–30]. Nitrogen for wheat production is equally important to realize the yield potential. Among the different elements of Bangladesh soil, N is the key input for achieving higher yield of wheat. Although application of nitrogenous fertilizer could increase yield to a certain level, it has adverse effect when the optimum level is exceeded [31]. Irrigation water dissolved the fertilizers and was made available to the crop for proper growth and development. Therefore, an attempt has been made to evaluate the effect of irrigation regimes and N levels and the best combination on the yield and yield contributing parameters of wheat.

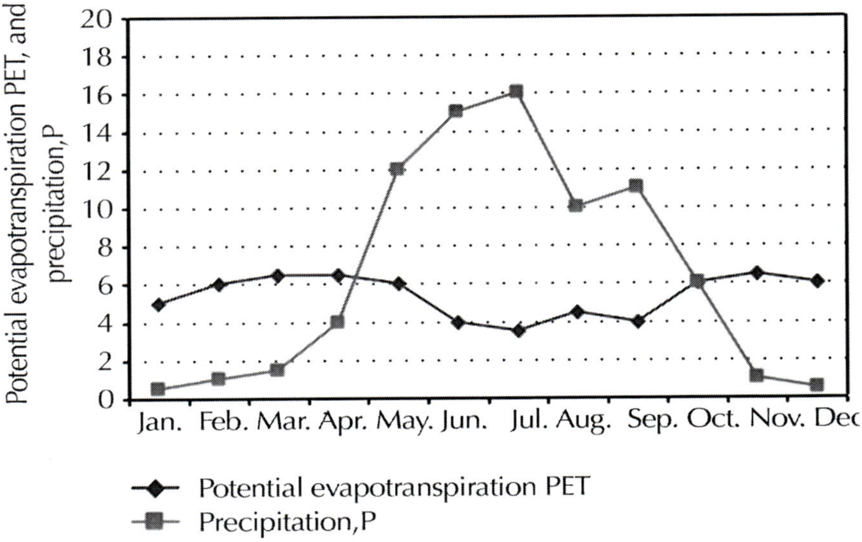

Figure 1: Hydrological balance of PET and precipitation at Mymensingh [28].

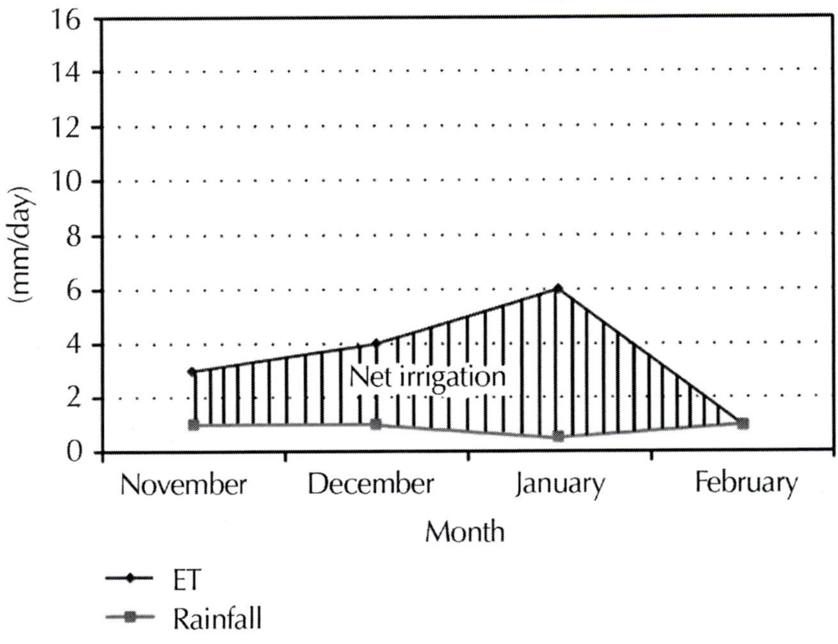

Figure 2: Net irrigation requirement of wheat [28].

MATERIALS AND METHODS

The experiment was conducted at Mymensingh region in Bangladesh using three farmer's fields to evaluate the effect of irrigation regimes and N level on the performance of wheat cv. Kanchan. The experiment included two factors (1) four irrigation regimes ((no irrigation), 100 mm at 30 days after sowing (DAS), 200 mm (100 mm at 30 DAS + 100 mm at 45 DAS), and 300 mm (100 mm at 30 DAS + 100 mm at 45 DAS + 100 mm at 60 DAS)) and (2) four N levels (00, 80, 100, and 120 kg N ha^{-1}). The experiment was laid out in a split plot design with irrigation as main plot and N as subplot. The soil texture is loam. During land preparation the soil pH ranged from 6.02 to 7.62 and organic carbon varied from 0.51to 1.63%. The land was prepared by the farmers with country plough followed by laddering. Three farmer's fields were selected for experimentation as replication. The unit plot size was 2.5 m × 4 m (10 m2). Triple superphosphate (TSP), muriate of potash (MP), and gypsum were applied at the rates of 80 kg P2O5, 60 kg K_2O, and 30 kg S ha^{-1}, respectively. One-third of the N along with the complete doze of TSP, MP, and S were applied at the time of final land preparation. The remaining two-thirds urea was top-dressed in two equal splits, namely, at 30 and 45 DAS. Data on grain and straw yield, plant height, number of ears per plant, length of ear, number of grains per ear, and 1000 grain weight were recorded. Data were analyzed following analysis of variance technique with the computer package M-stat and mean comparison was performed using Duncan's Multiple Range Test [32].

RESULTS AND DISCUSSION

Effect of irrigation regimes and N levels on the yield and yield contributing characters of wheat are presented in Table 1. The yield and yield parameters were significantly affected due to application of nitrogen and irrigation water. For higher yield, it is better to apply nitrogen at the stem elongation and heading unless it is not irrigated with minimum amount of water [33]. Maximum plant height was recorded in 300 mm irrigation treatment and shortest in the control. Availability of well distributed soil moisture at different growth stages due to irrigation probably enhanced the growth of plant. Due to application of irrigation water, ears per plant were significantly increased and followed similar pattern as in number of tillers per plant. Similar trend was found in the case of ear length. Plant height and ears per plant increased with the increasing rate of N doses but were not statistically different. Maximum number of tiller per plant was produced in 200 mm irrigation treatment which was statistically similar with other irrigation treatments except for the control. The effective tillers per plant were statistically different at 5% level of significance but identical among the nitrogen treatments. The maximum number of effective tillers per plant was obtained by applying 100 kg N ha^{-1} and the lowest in control. The result revealed that the applications of 80 to 120 kg N ha^{-1} treatment were statistically identical in respect of tillers per plant. Islam [34] reported that the maximum number of tillers per plant was obtained with 120 kg N ha^{-1} and the minimum in control. Ear length showed significant difference by the use of different levels of nitrogen. The longest ear was observed from application of 120 kg N ha^{-1} and the shortest in control. S. P. Singh and H. B. Singh [35] stated that ear length increased with increasing of nitrogen levels (0 to 120 kg N ha^{-1}). Grains per ear increased with increasing rate of nitrogen up to 100 kg N ha^{-1} and then decreased. The number of grains per ear was statistically similar due to application of 80 to 120 kg N ha^{-1} and significantly superior over control. Similarly results were obtained by Patel et al. [36]. Influence of irrigation on grain yield was statistically significant at 1% level of significance. Maximum grain yield was obtained in 200 mm irrigation treatment and minimum in control. These results are in conformity with the findings of Islam [34]. He reported that grain yield of 2.10 t ha^{-1} was obtained in 120 mm irrigation and 1.70 t ha^{-1} in control. The grain yields were significantly influenced by different levels of nitrogen. The grain yield was maximum due to application of 120 kg N ha^{-1} and statistically similar to that of 100 kg N ha^{-1} and both were significantly superior to 80 kg N ha^{-1}. These results are in agreement with the findings of Patel and Upadhyay [37]. Different irrigation regimes were found to have significant effect on the straw yield. Straw yield exhibited the tendency of increasing with the influence of irrigation levels. This might be due to the luxuriant vegetative growth in terms of plant height and number of tillers per plant. The maximum straw yield was obtained with 200 mm irrigation and minimum in control. Islam [34] reported that straw yield of 2.74 t ha^{-1} was found with the application of 180 mm irrigation water. Straw yield followed similar pattern to grain yield. This might be due to the exuberant vegetative growth noted in case of higher doses of nitrogen. Higher straw yield was obtained in 120 kg N ha^{-1} treatment and statistically superior to 0 to 80 kg N ha^{-1}. A significant variation was recorded for 1000 grain weight owing to difference in irrigation levels. 1000 grain weight was statistically similar in all irrigation treatments but the control produced the lowest weight. S. P. Singh and H. B. Singh [35] reported that 1000 grain weight varied significantly with variable number of irrigation. Different N levels had significant influence on 1000 grain weight. Maximum weight of 1000 grain was found with 120 kg N ha^{-1} and minimum weight was observed with control.

Interaction between irrigation and nitrogen did not show significant effect on plant height, tillers $plant^{-1}$, grains ear^{-1}, and 1000 grain weight. The interaction effect of irrigation and nitrogen application on ear length was statistically significant. The longest ear was observed due to 200 mm irrigation with 80 kg N ha^{-1}. Maximum grain yield was obtained due to application of 200 mm irrigation water and 120 kg N ha^{-1} (Figure 3) which was statistically similar to the 200 mm irrigation and 100 kg N ha^{-1} treatment. Minimum grain yield was obtained in the combination of no irrigation and no nitrogen treatment. These findings are corroborated with the observations by Kumar et al. [38]. Straw yields varied significantly at 1% level under different irrigation regimes and doses of nitrogen. It can be seen that straw yield increased with the increasing depth of irrigation and the highest rates of nitrogen fertilizer. Maximum straw yield was recorded by the application of 200 mm irrigation with 100 kg N ha^{-1} and the minimum straw yield was produced in control. Eventually, it may be concluded that a combination of 200 mm irrigation and 120 kg N ha^{-1} is the best in respect of yield and yield contributing characters.

Table 1: Effect of irrigation and nitrogen on the yield and yield contributing characters of wheat

Treatments	Plant height (cm)	Tillers plant^{-1} (no.)	Ears plant^{-1} (no.)	Ear length (cm)	Grains ear^{-1} (no.)	Grain yield (t ha^{-1})	Straw yield (t ha^{-1})	1000 grain weight (g)
Irrigation I_0 (00 mm)	62.75^{b}	2.49^{b}	2.18^{b}	7.35^{b}	35.75^{b}	1.74^{d}	4.12^{d}	42.83^{b}
I_1 (100 mm)	65.62^{ab}	2.85^{a}	2.53^{a}	7.78^{a}	36.82^{ab}	1.99^{c}	4.38^{c}	43.50^{ab}
I_2 (200 mm)	65.87^{ab}	2.90^{a}	2.70^{a}	8.04^{a}	38.23^{a}	2.27^{a}	5.12^{a}	44.57^{a}
I_3 (300 mm)	67.80^{a}	2.81^{a}	2.59^{a}	7.95^{a}	38.10^{a}	2.14^{b}	4.71^{b}	44.50^{a}
P	<0.05*	<0.01**	<0.01**	<0.01**	<0.01**	<0.01**	<0.01**	<0.01**
Nitrogen N_0 (00 kg/h)	63.92	2.63^{b}	2.40	7.02^{b}	34.15^{b}	1.84^{c}	3.96^{c}	42.58^{c}
N_1(80 kg/ha)	65.62	2.77^{a}	2.48	7.98^{a}	37.37^{a}	1.97^{b}	4.53^{b}	43.50^{bc}
N_2 (100 kg/ha)	64.65	2.83^{a}	2.53	7.88^{a}	38.92^{a}	2.15^{a}	4.87^{a}	44.58^{ab}
N_3 (120 kg/ha)	67.75	2.82^{a}	2.59	8.25^{a}	38.47^{a}	2.18^{a}	4.97^{a}	44.83^{a}
P	ns	<0.05**	ns	<0.01**	<0.01**	<0.01**	<0.01**	<0.01**

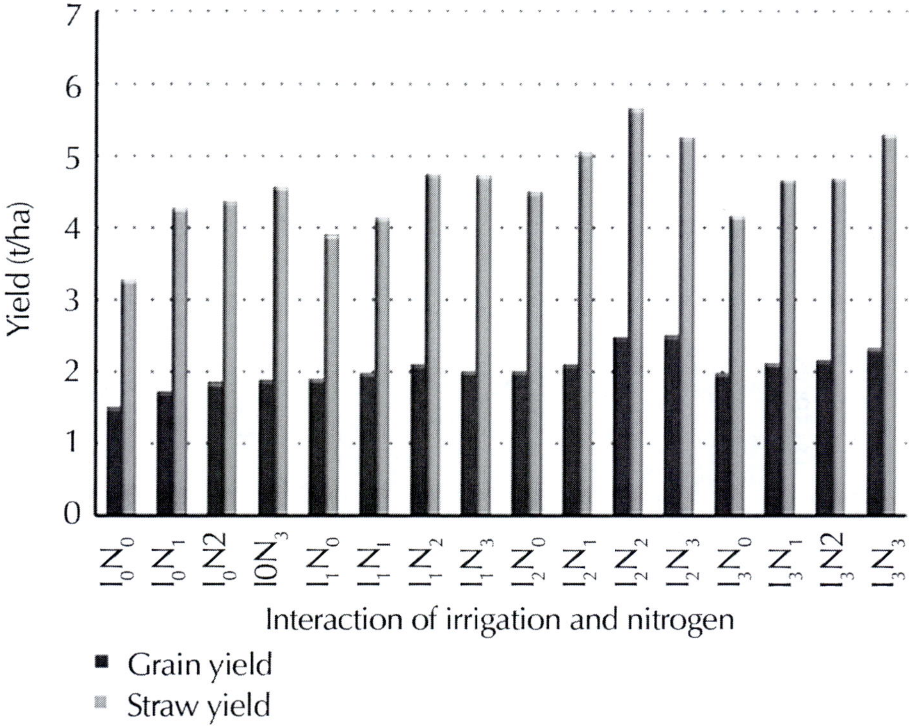

Figure 3: Combined effect of irrigation and nitrogen on the performance of wheat.

The moisture status of the experimental plots during growing period of wheat is shown in Figure 4. Percentage of soil moisture gradually decreased through the soil profile in control (Figure 4(a)). Moisture status at 20 and 40 cm depths of soil decreased days after sowing and then gradually increased at every time of irrigation applied (Figures 4(b)–4(d)). It might be due to elapsed time of percolation through the soil profile. Soil moisture status at 60 cm depths followed similar trend but moisture status at 20 cm depth fluctuated at prior and following irrigation, respectively. It might be due to infiltration, evaporation, and water holding capacity of the soil.

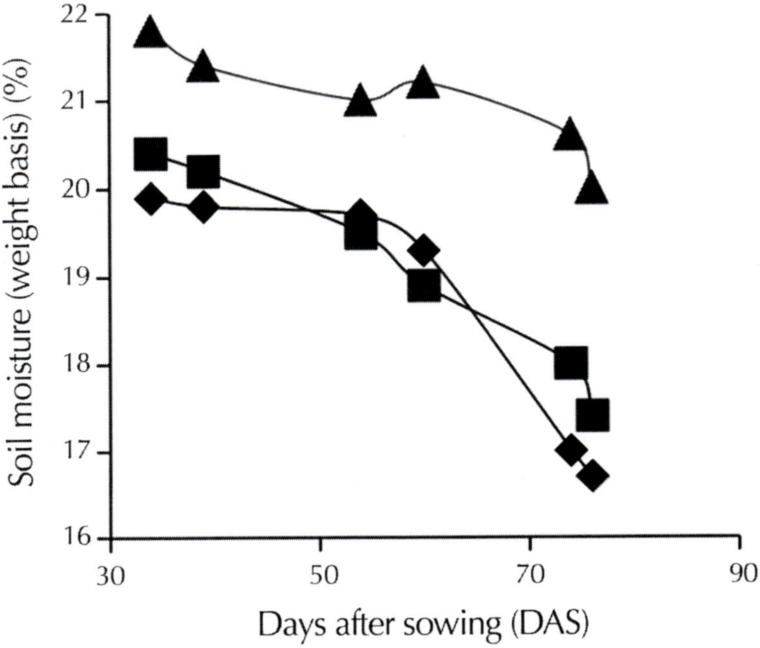

(a) Soil moisture status in control treatment

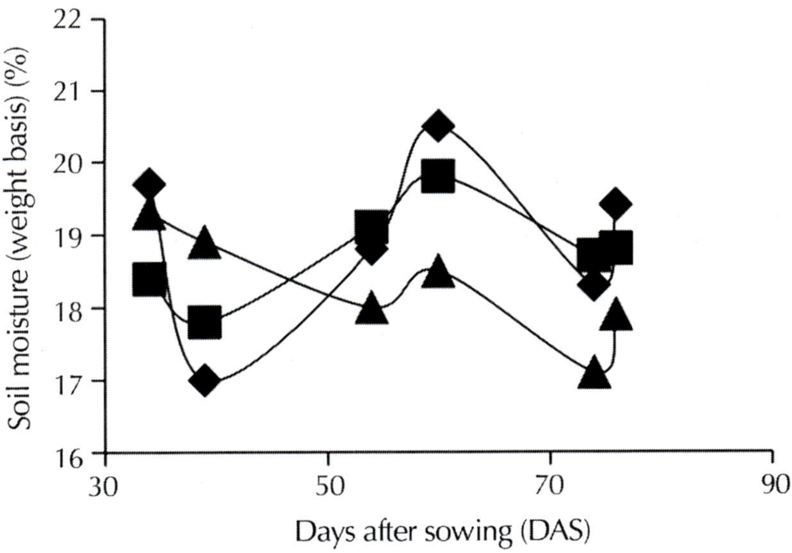

(b) Soil moisture status at 100mm irrigation treatment

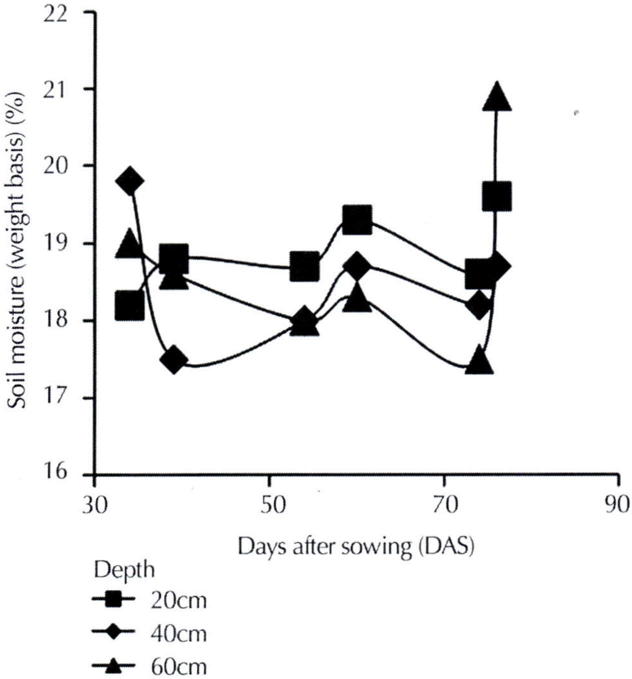

(c) Soil moisture status at 200mm irrigation treatment

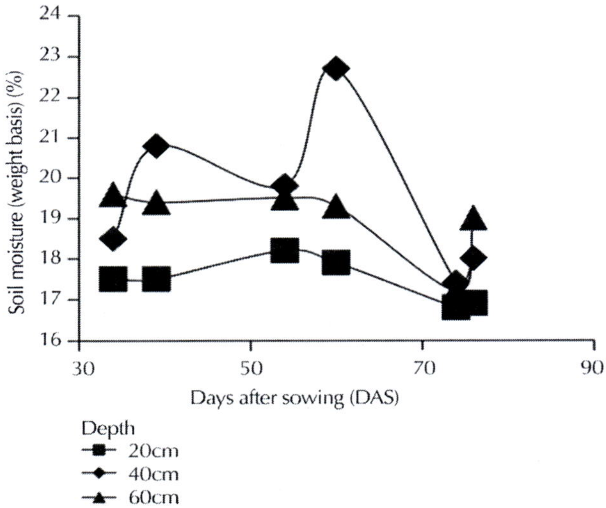

(d) Soil moisture status at 300mm irrigation treatment

Figure 4: Soil moisture status at different depth of soil in irrigation treatments.

The N contents in plots applied with 80, 100, and 120 kg N/ha treatments increased gradually up to 60 days after sowing (Figure 5). It may be due to the fact that one-third of the nitrogen was applied at sowing and the rest in two equal splits at 35 and 65 days after sowing (DAS). It can be seen that the nitrogen content decreased rapidly from 90 to 110 DAS because nitrogen was not applied at that period and also nitrogen uptake by plants may be higher due to flowering and grain formation.

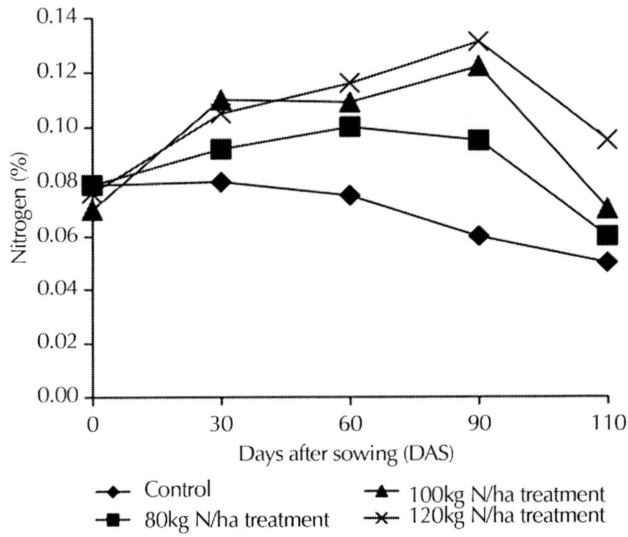

Figure 5: Nitrogen status of soil with different rates of N-application at different irrigation intervals (days).

CONCLUSIONS

The effects of combinations of irrigation regimes and nitrogen rates on yield and yield contributing parameters were studied at field level. Different doses of nitrogen significantly influenced the grain yield and yield parameters. For the highest grain yield, nitrogen doses of 100 kg N ha^{-1} were the best treatment when considering nitrogen fertilizer only. Irrigation regimes also have significant effect on yield and growth parameters of wheat. The combination of 200 mm irrigation and 120 kg N ha^{-1} is the best treatment for optimal production of wheat.

CONFLICT OF INTERESTS

The authors declare that there is no conflict of interests regarding the publication of this paper.

ACKNOWLEDGMENTS

Financial support by the research Grant no. PY/2014/01708 of Universiti Teknologi Malaysia (UTM) is gratefully acknowledged.

REFERENCES

1. BBS., Statistical Pocket Book of Bangladesh Bureau of Statistics, Statistic Division, Ministry of Planning, Government of the Republic of Bangladesh, 2007.
2. K. El Mejahed and L. Aouragh, "Green manure and n fertilizer in soil quality and profitability of wheat based system in semiarid morocco using nuclear techniques," in Management of Nutrients, Water in Rainfed Arid, Semi-Arid Areas for Increasing Crop, Production, IAEA-TECDOC-1468, pp. 89–106, 2005.
3. M. M. Rusan, A. Battikhi, and S. Zuraiqi, "Enhancement of nitrogen and water use efficiency by optimizing the combination of soil, crop and nitrogen management," in Management of Nutrients and Water in Rainfed Arid and Semi-arid Areas for Increasing Crop Production, IAEA-TECDOC-1468, pp. 155–177, 2005.
4. Z. Cui, X. Chen, and F. Zhang, "Current Nitrogen management status and measures to improve the intensive wheat-maize system in China," Ambio, vol. 39, no. 6, pp. 376–384, 2010.
5. V. K. Arora, H. Singh, and B. Singh, "Analyzing wheat productivity responses to climatic, irrigation and fertilizer-nitrogen regimes in a semi-arid sub-tropical environment using the CERES-Wheat model," Agricultural Water Management, vol. 94, no. 1–3, pp. 22–30, 2007.
6. H. Sun, C. Liu, X. Zhang, Y. Shen, and Y. Zhang, "Effects of irrigation on water balance, yield and WUE of winter wheat in

the North China Plain," Agricultural Water Management, vol. 85, no. 1-2, pp. 211–218, 2006.

7. S. Lenka and A. K. Singh, "Simulating interactive effect of irrigation and nitrogen on crop yield and water productivity in maize-wheat cropping system," Current Science, vol. 101, no. 11, pp. 1451–1461, 2011.
8. M. Montazar and M. Mohseni, "Influence of supplemental irrigation and applied nitrogen on wheat water productivity and yields," Journal of Agricultural Science, vol. 3, pp. 78–90, 2011.
9. F. Karam, R. Kabalan, J. Breidi, Y. Rouphael, and T. Oweis, "Yield and water-production functions of two durum wheat cultivars grown under different irrigation and nitrogen regimes," Agricultural Water Management, vol. 96, no. 4, pp. 603–615, 2009.
10. D. Shimshi and U. Kafkafi, "The Effect of supplemental irrigation and nitrogen fertilisation on wheat (Triticum aestivum L.)," Irrigation Science, vol. 1, no. 1, pp. 27–38, 1978.
11. R. H. Moll, E. J. Kamprath, and W. A. Jackson, "Analysis and interpretation of factors which contribute to efficiency of nitrogen utilization," Agron Journal, vol. 74, pp. 562–565, 1982.
12. M. A. Quayum and F. Kamal, "Effect of irrigation at different growth stages of wheat and grain yield," Bangladesh Journal of Agricultural Sciences, vol. 11, pp. 47–55, 1986.
13. K. D. Shing and B. M. Sharma, "Soil-test based specification for efficient use of fertilizer and targeted yield of wheat (Triticum aestivum) in Typic Ustochrept soil of Delhi Region," Indian Journal of Agricultural Sciences, vol. 60, pp. 36–40, 1990.
14. C. Cabeza, A. Kin, and J. F. Ledent, "Effect of water shortage on main shoot development and tillering of common and spelt wheat," Journal of Agronomy and Crop Science, vol. 170, no. 4, pp. 243–250, 1993.
15. G. Hussain and A. A. Al-Jaloud, "Effect of irrigation and nitrogen on water use efficiency of wheat in Saudi Arabia," Agricultural Water Management, vol. 27, no. 2, pp. 143–153, 1995.
16. H. Zhang and T. Oweis, "Water-yield relations and optimal irrigation scheduling of wheat in the Mediterranean region,"

Agricultural Water Management, vol. 38, no. 3, pp. 195–211, 1999.

17. G. Hussain, A. A. Al-Jaloud, and S. Karimulla, "Effect of treated effluent irrigation and nitrogen on yield and nitrogen use efficiency of wheat," Agricultural Water Management, vol. 30, no. 2, pp. 175–184, 1996.
18. R. Tavakkoli and T. Y. Oweis, "The role of supplemental irrigation and nitrogen in producing bread wheat in the highlands of Iran," Agricultural Water Management, vol. 65, no. 3, pp. 225–236, 2004.
19. S. M. Shirazi, S. Akib, F. A. Salman, U. J. Alengaram, and M. Jameel, "Agro-ecological aspects of groundwater utilization: a case study," Scientific Research and Essays, vol. 5, no. 18, pp. 2786–2795, 2010.
20. J. L. Cooper, "The effect of nitrogen fertilizer and irrigation frequency on a semi-dwarf wheat in south-east Australia. 1. Growth and yield," Australian Journal of Experimental Agriculture and Animal Husbandry, vol. 20, pp. 359–364, 1980.
21. P. R. Gajri, S. S. Prihar, and V. K. Arora, "Interdependence of nitrogen and irrigation effects on growth and input-use efficiencies in wheat," Field Crops Research, vol. 31, no. 1-2, pp. 71–86, 1993.
22. T. Kätterer, A. Hansson, and O. Andrén, "Wheat root biomass and nitrogen dynamics—effects of daily irrigation and fertilization," Plant and Soil, vol. 151, no. 1, pp. 21–30, 1993.
23. J. R. Frederick and J. J. Camberato, "Water and nitrogen effects on winter wheat in the southeastern coastal plain: I. Grain yield and Kernel traits," Agronomy Journal, vol. 87, no. 3, pp. 521–526, 1995.
24. S. Abderrazak, K. M. Ezzarouk, and M. Amuslim, "Effects of water stress and nitrogen fertilizer rate on nitrogen uptake, water use and wheat yield," Al Awamia, no. 89, pp. 49–75, 1995.
25. K. Singh and G. L. Jain, "Effect of sowing time, irrigation and nitrogen on grain yield and quality of durum wheat (Triticum durum)," Indian Journal of Agricultural Sciences, vol. 70, no. 8, pp. 532–533, 2000.

26. Z. Z. Xu, Z. W. Yu, D. Wang, and Y. L. Zhang, "Nitrogen accumulation and translocation for winter wheat under different irrigation regimes," Journal of Agronomy and Crop Science, vol. 191, no. 6, pp. 439–449, 2005.
27. B. D. Sharma, S. S. Cheema, and S. Kar, "Water and nitrogen uptake of wheat as related to nitrogen application rate and irrigation water regime," Fertilizer News, vol. 35, pp. 31–35, 1990.
28. S. M. Shirazi, Z. Ismail, S. Akib, M. Sholichin, and M. A. Islam, "Climatic parameters and net irrigation requirement of crops," International Journal of Physical Sciences, vol. 6, no. 1, pp. 15–26, 2011.
29. S. M. Shirazi, M. Sholichin, M. Jameel, S. Akib, and M. Azizi, "Effects of different irrigation regimes and nitrogenous fertilizer on yield and growth parameters of maize," International Journal of Physical Sciences, vol. 6, no. 4, pp. 677–683, 2011.
30. S. M. Shirazi, M. A. Islam, Z. Ismail, M. Jameel, U. J. Alengaram, and A. Mahrez, "Arsenic contamination of aquifers: a detailed investigation on irrigation and portability," Scientific Research and Essays, vol. 6, no. 5, pp. 1089–1100, 2011.
31. N. Kataria and K. Bassi, "Effect of organic mulch and nitrogen on early-sown wheat (Triticum aestivum) under rainfed conditions," Indian Journal of Agronomy, vol. 42, no. 1, pp. 94–97, 1997.
32. K. A. Gomez and A. A. Gomez, Statistical Procedure for Agricultureal Research, John Willey & Sons, New York, NY, USA, 2nd edition, 1984.
33. C. G. Abourached, S. K. Yau, M. N. Nimah, and I. I. Bashour, "Deficit irrigation and split N fertilization on wheat and barley yields in a semi-arid mediterranean area," The Open Agriculture Journal, vol. 2, pp. 28–34, 2008.
34. M. S. Islam, Growth and yield response of wheat to irrigation and nitrogen application [M.S. Thesis], Department of Soil Science, Bangladesh Agricultural University, Mymensingh, Bangladesh, 1997.
35. S. P. Singh and H. B. Singh, "Effect of irrigation time and nitrogen level on wheat (Triticum aestivum) under late-sown condition of Western Uttar Pradesh," Indian Journal of Agronomy, vol. 36, pp. 41–42, 1991.

36. N. M. Patel, S. G. Sadaria, B. B. Kaneria, and V. D. Khanpura, "Effect of irrigation, potassium and zinc on growth and yield of wheat (Triticum aestivum)," Indian Journal of Agronomy, vol. 40, pp. 290–292, 1995.
37. R. M. Patel and P. N. Upadhyay, "Response of wheat (Triticum aestivum) to irrigation under varying levels of nitrogen and phosphorus," Indian Journal of Agronomy, vol. 38, pp. 113–115, 1993.
38. Kumar, D. K. Sharma, and H. C. Sharma, "Response of wheat (Triticum aestivum) to irrigation and nitrogen in sodic soils," Indian Journal of Agronomy, vol. 40, no. 1, pp. 338–342, 1995.

Chapter 9

Optical Sensors Applied in Agricultural Crops

Fabrício Pinheiro Povh[1] and Wagner de Paula Gusmão dos Anjos[1]

[1]Fundação ABC, Agricultural Machinery and Precision Agriculture Department, Castro, Brazil

INTRODUCTION

There is a very wide range of optical sensors applied in agriculture, which goes from sensors used to analyze soil attributes to sensors installed in combines to measure protein content in wheat grains while they are being harvested. But in this chapter we are going to discuss about optical sensors that, from a short distance are able to measure the agricultural crop's reflectance using specific wavelengths, and how can we use this information.

This kind of optical sensors began to be studied in 1991 with the development of sensors focused in weed detection at the Oklahoma State University. Just based on the simple fact that soil and plants (weeds) have a different interaction with the light emitted from the sensors, allowing identifying what is soil and what is a plant.

In 1992 there was the first discussion between the Departments of Plant and Soil Sciences and Biosystems and Agricultural Engineering concerning the possibility of sensing biomass in wheat and bermudagrass. The objective was to use biomass as an indicator of nutrient need (based on removal). In 1993 Dr. John Solie, Dr. Marvin Stone and Shannon Osbourne collected sensor readings at ongoing bermudagrass, with nitrogen rates versus nitrogen timing experiments with the Noble Foundation in Ardmore, Oklahoma. Initial results were promising enough to continue this work in wheat. And in fall of 1993 the first variable application of nitrogen was done across a 70 meter transect. In 1994 John Ringer and Shannon Osbourne collected sensor readings and later applied variable N fertilizer rates based on the first bermudagrass algorithm developed by TEAM-VRT.

In the subsequence years the research advanced creating algorithms for nitrogen application in many different crops. And nowadays we have commercial sensors being sold to farmers to make real-time application of nitrogen, growth-regulators and desiccants. The objective of this chapter is to show different applications of optical sensors in agricultural crops. At the ABC Foundation (A private research institution maintained by farmers created in 1984) located in south Brazil, the studies began in 2006 with commercial sensors. Making applied research, the results can be delivered immediately to the agronomists of five cooperatives (Capal, Batavo, Castrolanda, Coopagrícola and Holambra).

The topics to be discussed in this chapter are:

- The types of sensors used;
- The range in the spectrum used by sensors and response on crops;
- Aspects of use in the field and what could affect the measurements;
- Experiments and the results obtained with sensors;
- Use of optical sensors combined with a GPS receiver to create maps;
- Nitrogen application based on sensor's measurements;

THE TYPES OF SENSORS USED

The sensors to measure crops reflectance can be classified according to the platform, like satellites, aerial (airplanes, UAV's - unmanned aerial vehicles, balloon) and ground based. For satellites, airplanes and UAV's it is most common to use cameras to collect images, and ground based optical sensors can collect reflectance data and storage in a text file. The ground sensors can also be classified into active or passive. The basic difference is that the passive sensors need an external source of light, like the sun. The active sensors have its own source of light, which can be a wide range light or a specific wavelength.

There are available a few brands on the market, each one with its own construction characteristics like internal batteries, GPS antenna, data logger and log frequency. The Table 1 has some examples:

Table 1: Optical sensors and manufacturers

Manufacturer	Sensor	Country
AgLeader	OptRx	United States
Falker	ClorofiLOG	Brazil
Force A	Dualex and Multiplex	France
Fritzmeier	ISARIA	Germany
Holland Scientific	CropCircle	United States
Konica Minolta	SPAD	Japan
Topcon	CropSpec	Japan
Trimble	GreenSeeker	United States
Yara	N-Sensor	Germany

There are also some differences about the distance from the sensor to the crops. For example, the SPAD and the ClorofiLOG sensors need to make static measurements, touching the sensors on crop's leaves. The other sensors make the measurements from centimeters to meters, but do not need to have contact with the leaves.

Figure 1: Crop Circle optical sensor.

Let's use one of the available sensors as an example, the Crop Circle® ACS-210 sensor, manufactured by the Holland Scientific Inc., Lincoln, Nebraska. As show in the picture (Figure 1) this sensor has one LED (Light Emitting Diode) that emits active radiation simultaneously in visible and near infrared light with a system called PolySource™ and two silicon photodiodes with a spectral range of 320 to 1,100 nm to detect light. One detector works between 400 and 680 nm and the other between 800 and 1,100 nm. Using a filter on each detector, the wavelengths used are the amber light (590 ± 5 nm) and near infrared (880 ± 10 nm). The sensor must be placed between 0.25 and 2.13 m from the target, and then the light reaches the target and reflects part of the energy, which is received by the detectors. This sensor needs external batteries (12 V) and GPS antenna, but has its own data logger, that saves the data in a SD card. Also in Figure 1 there is an example on how to install the sensors in a motorcycle to collect data on-the-go.

The Figure 2 has two other examples, one is the Hand held GreenSeeker, from Trimble, and the other one is a prototype from Oklahoma State University, which now is also being commercialized by Trimble. The Pocket Sensor was designed for small farmers, made to be a low cost sensor, it does not use GPS signal or a data logger, and you can just position the sensor over the crop, press the trigger and see the value of NDVI (Normalized Difference Vegetation Index).

Figure 2: GreenSeeker sensor collecting NDVI values for weed (a) and (b); detailed pictures from GreenSeeker Hand Held (c) and Pocket Sensor (d); and GreenSeeker mounted on a sprayer (e) and on a motorcycle (f) collecting NDVI values in sugarcane.

The Range in the Spectrum Used by Sensors and Response on Crops

The electromagnetic spectrum goes from gamma rays to radio waves. These sensors used for measure crop reflectance usually work in the visible and near infrared region of the spectrum, and combining at least two wavelengths to calculate vegetation indices. From the agronomic point of view, the visible light has a straight relationship with the chlorophyll content, absorbing the blue and red lights, and reflecting the green light. That's what makes us to see healthy plants as green. The near infrared light, not visible by the human eye, is reflected by the mesophyll cells, which is found in more quantity in a plant than chlorophyll, resulting in a much higher reflectance than visible lights. Using both wavelengths it is possible to evaluate the color and biomass production of a crop. In practice, greener and higher biomass plants have a higher chance to have higher yields.

Remote sensing can be defined as the technique of acquisition and application of information about an object without any physical contact. The information is acquired by the detection and measurement of changes that an object imposes to the environment around it, and this signal may include an electromagnetic field emitted and/or reflected, acoustic waves reflected and/or disturbed by the object or the disturbances of gravitational field or magnetic potential with the presence of the object. Usually the acquisition of information is based on capturing the electromagnetic signals that cover the entire spectrum of electromagnetic waves to radio long waves, passing through microwaves, submillimeter, thermal and near infrared, visible, ultraviolet, X ray and gamma ray [1].

The sensors used by remote sensing are devices capable of detect and register electromagnetic radiation in certain range of electromagnetic spectrum and generate information that may be transformed in a product, which can be interpreted, like an image, graphic or tables. The sensor systems are basically formed by an optical part, constructed with lens or mirrors that have the objective of receiving and directing the energy from the targets to the detectors.

The spectral reflectance measurements are the non-destructive approach most promising for determining nitrogen deficiency in crops [2]. For this purpose, remote sensing has been used to evaluate crop conditions related to nitrogen [3]. Many researchers used remote sensing to estimate crop parameters like LAI (Leaf Area Index) [4], leaf chlorophyll content [2], soil cover [5], dry matter [6], water content [7], yield [8], nitrogen content [9] and many others.

In reference [10] studying spectral signature of green leaves, was found that wavelengths from 400 to 700 nm (visible), the reflectance is low, about 10%, with a smooth increase at 550 nm (green). In the near infrared region (700 to 1,300 nm) there is another increase in the reflectance, close to 50%. For the visible light the low reflectance is related to the absorption associated to leaf pigments, mainly chlorophyll. And the reflectance increase at the near infrared region is due the internal structure of the leaves (size and shape of the cells and empty spaces). Combining both visible and near infrared reflectance there are the vegetation indices, which can be resulted from two or more spectral bands.

First, [11] it was proposed the ratio between the measurements from 800 and 675 nm to determine the leaf area index in forests. The relation between these two wavelengths is known as RVI (Ratio Vegetation Index). The NDVI (Normalized Difference Vegetation Index) appeared just after [12], which was found a relation between two wavelengths that better solved the issues about soil interference on the vegetation measurements, and also reduced the atmosphere influence and sun angle variations.

The normalization proposed guarantees that the values obtained from the NDVI are contained in the same scale of values, between -1 and 1, as shown in the equation (1).

$$NDVI = \frac{(\rho_{IR} - \rho_{V})}{\rho_{IR} + \rho_{V}} \quad (1)$$

Which:

ρ_{IR} ρIR = infrared reflectance;

ρ_{V} ρV = visible reflectance.

The normalization is produced by the combination of the strong absorption by the chlorophyll in the red region and the strong reflectance in the near infrared, due to the dispersion in the leaf mesophyll and the absence of absorption by the pigments [13]. The peculiarity assigned to the NDVI is the early saturation, what makes it insensible to biomass increase after certain development stage. That means the NDVI stabilizes, showing constant values, even with the biomass increase [10].

ASPECTS OF USE IN THE FIELD AND WHAT CAN AFFECT THE MEASUREMENTS

Measurements in the field can be affected by the sensor positioning, like the distance from the crop, dependence of a light source, the presence of dew over the leaves and also because of factors that can stress the plants. The light source classifies the sensors in active and passive sensors. The passive sensors are dependent of sunlight, not working at night or might show different readings when there are clouds or shadows. There is a limitation of the distance from the target to the sensor, because if it is too close the sensor may not capture the reflectance, and if it is too far the data may have noise signals. And the presence of dew is just because the presence of water over the leaves can change the reflectance in both visible and near infrared. With the presence of dew the reflectance increases, but as visible light is more affected, consequently reduces the NDVI values.

Other methods can be used instead of on-the-go sensing to indirectly measure nitrogen stress, like chlorophyll content in the leaves using chlorophyll meters [14] like SPAD and ClorofiLOG. In reference [15], analyzing aerial and satellite images it was found lower correlation with wheat crops variables then ground sensors, besides depending on the weather it is not possible to have satellite images due the presence of clouds.

The spectral data that can be obtained from satellite images have low temporal resolution to use in agriculture. The ground acquisition is independent from the weather, the data collection can be done together with other machinery operation and the data are available just after you finish, with no need of complicated processing [16]. Higher correlation was found between the NDVI obtained from the

GreenSeeker then the NDVI obtained from the satellite Quickbird II images for the amount of nitrogen applied, nitrogen content on the flag leaf, yield and protein content in wheat grains [15].

In reference [17] the author used a sensor installed in the front of a tractor to acquire spectral data in wheat. The tractor was at the speed of 0.8 m/s, and the sensor was set to register data with the frequency of 10 measurements per second. The authors found good relationship between the reflectance and nitrogen absorption by the wheat crop. The same configuration was used to collect data in pasture and also found good relationship between the NDVI and the nitrogen removal by the crop [18]. But depending on the sensor used, the speed can also affect the NDVI, increasing the coefficient of variation of the measurements.

An experiment conducted along the day, collecting NDVI values at the same spot but in different times of the day showed that the presence of dew on the leaves reduced the NDVI values of 12% for GreenSeeker and 27% for Crop Circle, from the first (7:30 am) to the last reading (11:30 am). Near infrared showed small differences in the readings, not significant, unlike the results of the visible data, but they both increased. These results lead to conclude that the visible wavelength is more affected by the presence of dew than the near infrared. Then, the presence of dew should be considered for nitrogen recommendation based on active optical sensors.

Stressed plants can show a decrease in the absorption by the chlorophyll, also decreasing the reflectance in the near infrared due to changes in the cells structure [19]. And that is what makes the sensors to be promising while evaluating nitrogen stress. Many studies were realized to estimate nitrogen deficiency in corn [3], wheat [20], edible beans [21], cotton [22], citrus [23], barley [24] and sugarcane [25]. These authors showed great potential to use sensors to estimate nitrogen content in crops.

EXPERIMENTS AND RESULTS OBTAINED WITH SENSORS

As discussed previously, the sensors are able to measure differences in color and biomass production. In the field when nitrogen is applied and the plant absorbs it, the result is a plant greener and with more biomass. The Figure 3 shows the first experiment realized by ABC Foundation in Brazil in the winter of 2006 with the result of nitrogen application in wheat and the effect on the NDVI measured by the GreenSeeker sensor in two different soils.

The graphic in Figure 3 is a typical result of nutrient application in agricultural crops, with a quadratic model. There is always a limit, technical or economic, which beyond that there is no gain in yield. If linear models explained fertilizers application, where more fertilizer applied should result in more yield, it would be much easier, but there is also a physiological response, when higher nutrients rates become toxic to the plant. This result shows that applying more nitrogen we could find a response from the NDVI measured with a GreenSeeker sensor. In Table 2 it is possible to see other variables that had significant correlation with the NDVI in the same experiment.

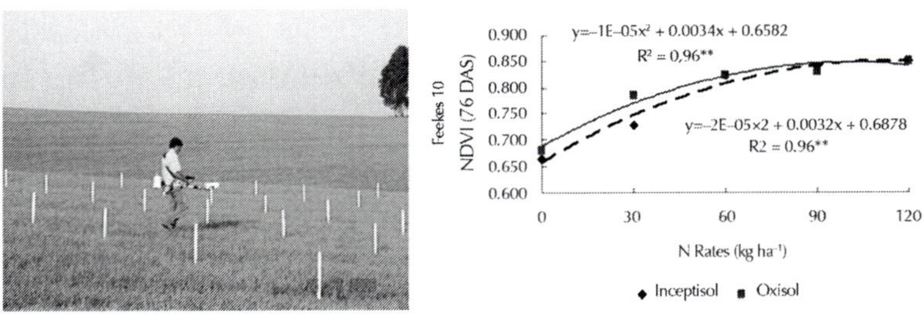

Figure 3: First experiment and NDVI response with nitrogen rates in wheat.

Table 2: Coefficients of correlation (r) between the evaluated variables in a wheat experiment in two different soil types. Leaf nitrogen content, dry matter and NDVI were evaluated 76 days after sowing, at the Feekes 10 stage of development

Variables	**Inceptsol**	**Oxisol**
	(r)	
Leaf N x Yield	0,99**	0,95**
Dry Matter x Yield	0,98**	0,97**
NDVI x Leaf N	0,95**	0,88*
NDVI x Dry Matter	0,97**	0,99**
NDVI x Yield	0,95**	0,98**

Use of optical sensors combined with a GPS receiver to create maps

Maybe one of the most interesting applications for optical sensors in agriculture is to be able to use geographical coordinates to create maps from reflectance measurements. But why do that? Precision Agriculture is based on the fact that all fields have variability. Understanding this variability is first step to make decisions about investments in Precision Agriculture. Many procedures may be used to characterize and treat spatial variability on yield aiming profit for the farmers, but a wide and safe vision about the impact of the variability in a production system requires an accurate measurement of this variability.

Soil variability is caused by climate, topography, vegetation, soil forming processes and also management. These factors can influence the variability in different scales and cause great variability on nutrient availability in the soil. Then, when using uniform rates of fertilizers it is almost certain that excessive rates will be applied to some parts of the field and inadequate rates in others.

Precision Agriculture can be defined as a management system that considers spatial variability that is present in a production field, independently of the size and treats locally this variability. It is well proved that quality and yield are spatially variable and systems are being developed to explore these variations and increase profit [26]. The variable rate application of fertilizers is one of the options to manage variability, and creating maps from optical sensors can help to realize variable applications with nitrogen.

An optical sensor connected to a GNSS (Global Navigation Satellite System) receiver is able to register reflectance values with a pair of coordinates (latitude and longitude), and when the file is imported to a GIS software (Geographic Information System) it is possible to represent the measurements and its distribution in a field with maps (Figure 4).

Depending on sensor system used, that information can be stored in different ways and types of files. Some systems installed in agricultural machinery allow using a laptop with software that will store the sensor readings and coordinates. Other systems use their own data logger or a pocket pc. The Figure 4is an example of a raw data collected in a field. Each strip is around 24 meters from each other, but what looks like a line, actually are many dots very close to each other due the frequency used to collect the data in the field. If we look closer, we see as in Figure 5.

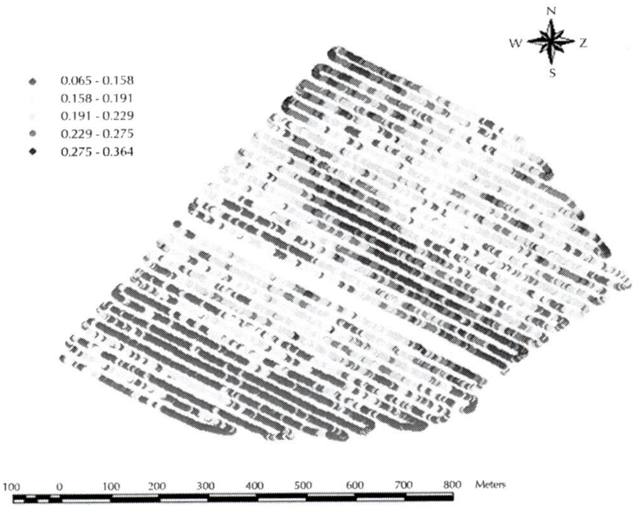

Figure 4: NDVI raw data.

Figure 5: NDVI raw data (zoom in).

As showed before in the picture of the motorcycle with one sensor on each side, when looking closer at the data we can see that in every coordinate we have a NDVI value from each sensor. As the frequency is very high, lets say 5 Hz or 5 measurements per second, if the motorcycle moves at 5 m/s (18 km/h) with 24 m distant from the next pass, the sensor registers two NDVI values every meter and more than 800 measurements per hectare. But the raw data may have noise, so after applying some statistical filters to remove outliers and using some kind of interpolation method, for example the inverse distance, it is possible to create a surface map with a regular grid that will look like a raster image. The Figure 6 is a NDVI map after interpolating the raw data from a corn field.

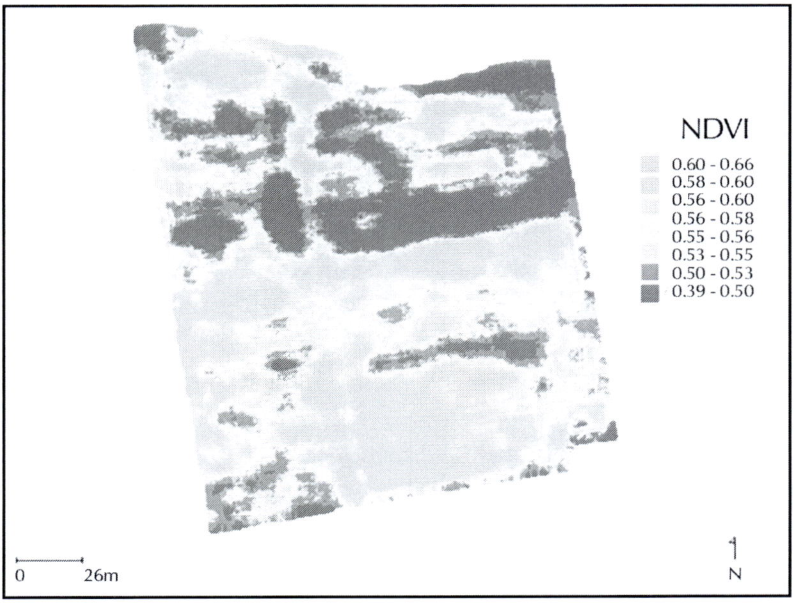

Figure 6: Interpolated NDVI map of a 2.86 ha field with a corn crop.

Figure 7: Picture of corn plants with low NDVI values.

This maps shows that in the red areas the corn plants had a low development and because of that will have lower yields (less than 10 tons/ha), while the green areas had yields over 12 tons/ha. This information is important for precision agriculture users that are interested in finding the specific problem for that particular location. Looking at plants in the field, right in middle of the map in the red area, we can see the plants as showed in Figure 7.

The NDVI map created while the corn crop is still at the V6 stage of development, allow the farmers and agronomists to go back in the field and see what is causing this slower growth of the plants. And as the information is stored, it is possible to go back every year at the same spot for comparison, collect soil and plant samples, send to the lab and try to understand what the problem is. That means the farmer can treat each part of the field accordingly with its own characteristics.

Nitrogen application based on sensor's measurements

Why manage nitrogen in agricultural crops? Because some crops do not have the ability to fix atmospheric nitrogen like soybeans do, so nitrogen is applied to crops from mineral fertilizers. Among the nutrients, nitrogen (N) is the most important and essential for crop development and is also important from an environmental perspective. Nitrogen is a constituent of chlorophyll, the first pigment to absorb the light energy necessary for photosynthesis. Plants usually have 1 to 5% nitrogen in their tissues, and if used appropriately with the other nutrients, N addition can accelerate the development of corn and other grains [27]. And unlike other important nutrients as phosphorus or potassium that are measured in the soil with chemical analyses of soil samples, nitrogen is recommended based on historical response curves.

Plants with N deficiency have yellow leaves and reduced growth. Not only does the absence of N limit yield, but so can excess N. In most cases, N application recommendations are based on average conditions, and farmers do not use the most advantageous fertilizer combinations, even when there is no financial limitation [28].

The absorption of N by crops is variable among and between seasons, as well as between locations in the same field, even when the N supplies are high [29]. The N supply from soil to crop varies spatially. Consequently, the demand for N by the crop also varies. As a result, the crop's nutritional status is a good indicator of the necessary N rate application [30].

Traditional N management around the world is generally inefficient, creates environmental contamination and is controversial. To provide appropriate recommendations of spatial N applications, it is necessary to use several tools simultaneously, such as crop and soil sensors, to achieve reliable measurements of N availability from soil and crops needs [31].

The evaluation of nitrogen use efficiency (NUE) in agriculture is an important way to evaluate the density of N applied and its role in yield [32]. Because crop responses to N application depend on the organic matter in the soil, strategies of N management in cereal crops that include reliable predictions of the response index in each season could increase NUE [33].

In this scenario, sensors are becoming more prevalent in agricultural lands. Using variable rate equipment, it is possible to detect variability in crops and make rapid decisions in the field. Some sensors allow real time changes in agricultural practices by detecting variability and responding to that variability [34].

In reference [20] they developed a methodology to apply nitrogen based on crops reflectance, using a yield parameter sensible to local conditions, intrinsic and that could reflect yield potential, possible to be used in season. Unlike other models that need several parameters to predict plant growth, the optical sensors use the plant as an indicator. Later [35] they created an index called INSEY (in-season estimated yield), which is calculated dividing the NDVI value by the number of days from sowing to sensing. The relation between the INSEY and Yield generate an exponential model using plot experiments, which can be used to estimate yield just based on the sensor measurements. Estimating yield where there is and where there is not nitrogen, by the difference in yield it is possible to calculate the nitrogen rate to make them to have the same yield.

This method considers the spatial variability of yield potential, the absorption of nitrogen applied at sowing and the crop's response to an additional N rate. NUE were increase in 15% and showed that the measurement of crop's reflectance can be used to calculate more efficient and profitable N rates [35]. The ability to identify areas where the crop will respond to the fertilizer applied is important, if the response to N is expected, then management strategies can be modified [36].

Many field experiments were realized in order to create the algorithms that convert the sensor readings in nitrogen rates and to validate the results. Depending on the year, savings of none to 75% of nitrogen were found. For example the Figure 8 shows the NDVI and yield response to nitrogen rates. Using the 120 kg of nitrogen per hectare as the standard rate, the NDVI measured 50 days before the harvest indicates that the wheat crop won't respond to rates over than 60 kg of nitrogen per hectare. And the wheat yield obtained after the harvest shows that rates over than 60 kg/ha will affect yield negatively. Therefore, in this example, the nitrogen rate was 50% lower than the commonly recommended rate.

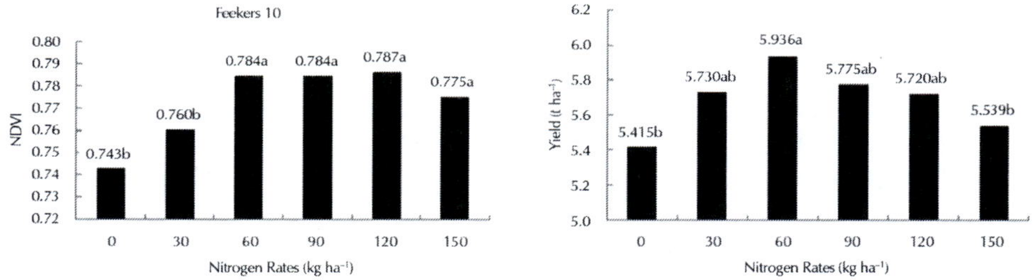

Figure 8: NDVI and wheat yield response with nitrogen rates applied.

The yield goal, often used for nitrogen recommendations, is the yield that the farmer expects to produce. But what you expect to produce and what really is going to be produced are usually different. In [20] they exemplify that the North Dakota State University recommend using the maximum yield obtained in the last 5 years as yield goal, because it is usually 30% higher than the average. The yield goal can also be calculated adding 5 to 10% to the average yield from the last 5 to 7 years [37]. One example is to use 25 kg of nitrogen per ton of corn expected. But management practices and the weather have a huge influence on yield. Climate can vary significantly from one year to another, what may cause great differences in yield potential [31].

Now let's use an experiment as an example of how to realize a nitrogen recommendation using variable rate across a field. The experiment was realized in South Brazil, Paraná State, farm Manzanilha (25° 28′ 55″ S, 50° 21′ 07″ W), during 2007 winter. The sensor used was the GreenSeeker Hand Held™ measuring the NDVI. The treatments (Table 3) were disposed in strips with fixed rate of nitrogen (6 x 600 m) and variable rate of nitrogen (11 x 600 m), with four treatments and four replications. The wheat was sowed in May 28th. The Sketch is presented in Figure 9.

Table 3: Treatments with nitrogen rates (kg/ha)

Treatments	N Sowing	N Feekes 1	N Feekes 5	N Feekes 10	Total N
1	18,4	0	0	Sensor	18,4
2	18,4	34	0	Sensor	52,4
3	18,4	0	102	0	120
4	18,4	34	68	0	120

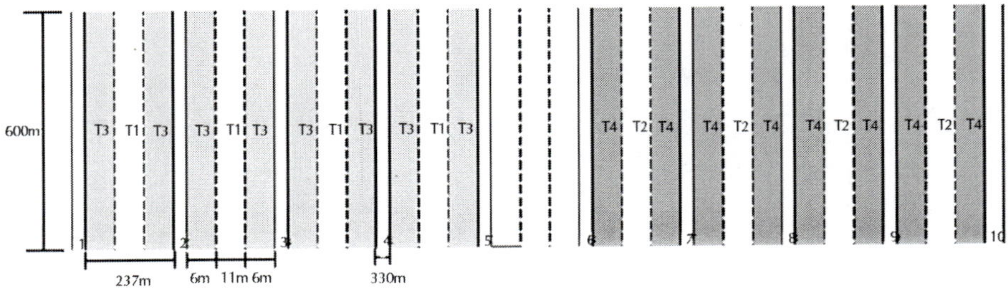

Figure 9: Experiment sketch.

All treatments received 18.4 kg/ha of nitrogen at the sowing; treatments 2 and 4 had an extra 34 kg/ha just after sowing and treatments 3 and 4 received more 102 and 68 kg/ha of N, respectively, on Feekes 5 to complete the 120 kg/ha of N, recommended based on historical response curves. Treatments 3 and 4 were used as control, and treatments 1 and 2 received variable rates of nitrogen at Feekes 10 based on the sensor readings.

The measurements were realized with the GreenSeeker sensor mounted on a motorcycle 79 days after sowing (Feekes 10). The pictures of the field and data collection can be seen in Figure 10.

(a) (b)

Figure 10: Details of the sensor mounted on the motorcycle (a) and data collection (b).

After the measurements, the data were imported to a GIS software (SSToolBox) to create a map with the NDVI raw data (Figure 11).

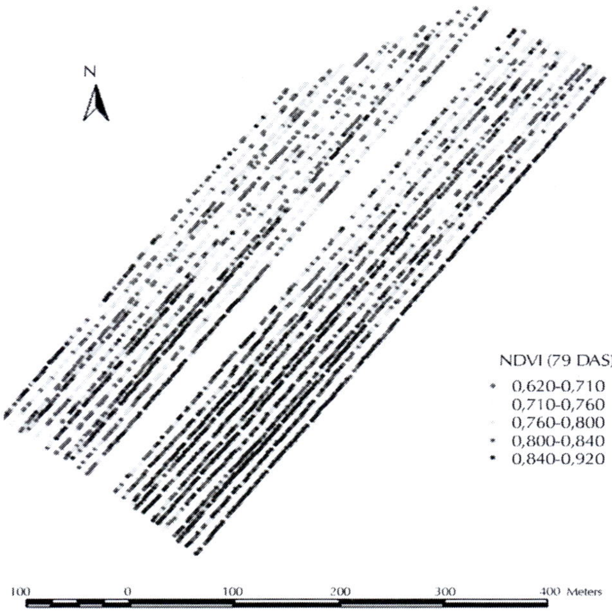

Figure 11: NDVI raw data in wheat.

Looking at the NDVI map it is possible to notice that even at the 120 kg/ha of N strips there was spatial variability, so it was decided to adapt the methodology proposed in [35], and do not use the average of the strip as a control, but divide the strip in three levels according with the topography. The elevation map (Figure 12) was created from the data collected by the GPS receiver. It was used three NDVI averages for treatment 3 and more three averages for treatment 4, with a total of six NDVI values to be used as a reference and estimate yield.

After dividing the elevation in three ranges it was calculated the average of the NDVI values for the 6 regions. Then the response index (RI) was calculated (eq. 2) with the ratio between the average NDVI from strips with 120 kg/ha of N and the NDVI from every point in the map. That means that every point has an NDVI value and a RI value.

$$RI = NDVI_{rich\ strip} / NDVI_{field} \qquad (2)$$

Which:

$NDVI_{rich\ strip}$ = NDVI from 120 kg/ha of N strips;

$NDVI_{field}$ = NDVI from all points in the field that will be applied N;

RI = response index.

The response index means that if the number is higher than 1, the NDVI from the strip with more N is higher than the NDVI from the parts of the field that will be applied, so the plants are using the N. And if RI = 1 means that where there is N and where there is no N the NDVI is the same, so there is no need to apply more N. Using RI it is possible to identify the parts of the field that will need more nitrogen, and save where there is no response, because it will not have increase in yield.

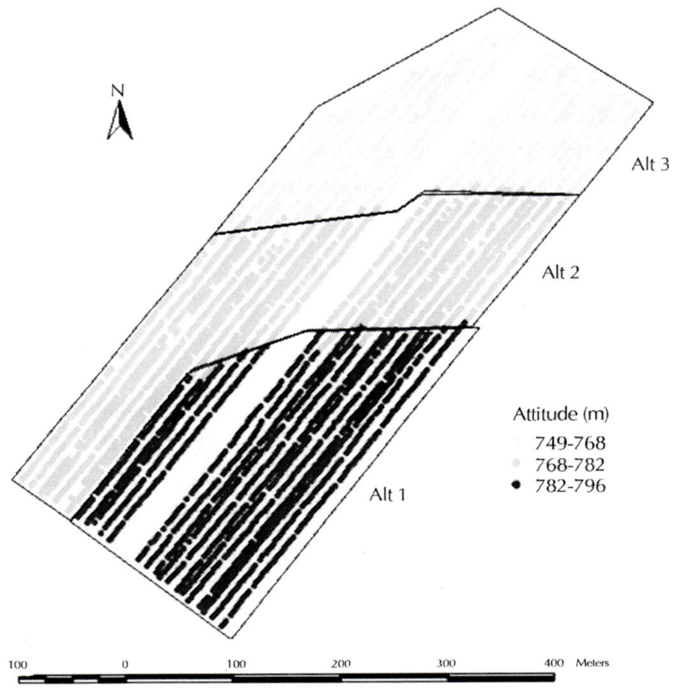

Figure 12: Elevation map divided in three levels.

To calculate the N rate was used the methodology from [35], identifying the difference of yield between the N rich strip and the rest of the field, and then applying the correct N rate to reach the same yield. That means the objective is to save nitrogen and not increase yield. The N rates calculated were used to create a nitrogen application map (Figure 13).

The calculated rates varied from 0 to 60 kg/ha of N, but they were simplified into three ranges and applied the maximum of each range. For example, the rate 20 kg/ha represents the rates from 0 to 20 kg/ha and so on. This simplification was done to make the application easier. To apply the nitrogen it was used a sprayer mounted on a tractor for liquid nitrogen application (Figure 14). To vary the rates along the field the tractor changed the speed, for each one of the rates was determined a specific gear of the tractor and a specific engine rotation. The sprayer was calibrated to apply 20 kg/ha (2.5 m/s), 40 kg/ha (1.25 m/s) and 60 kg/ha (0.8 m/s). The gears and the engine rotation for tractor were determined with a 50 m space measuring the time that the tractor took to move the 50 m. The sketch of the application is in Figure 15.

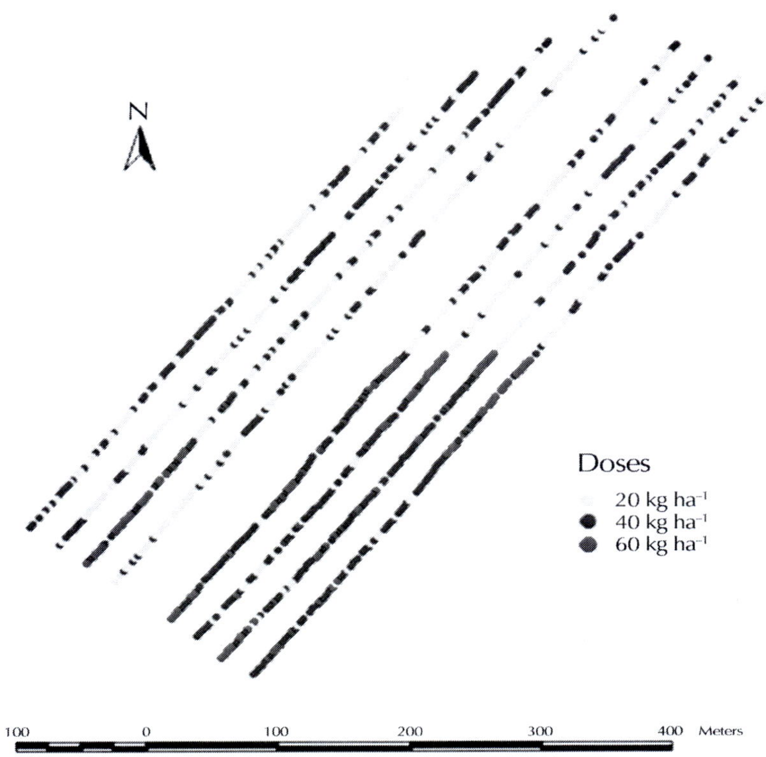

Figure 13: Nitrogen map from the calculated rates.

Figure 14: Pictures of the liquid nitrogen fertilizer application.

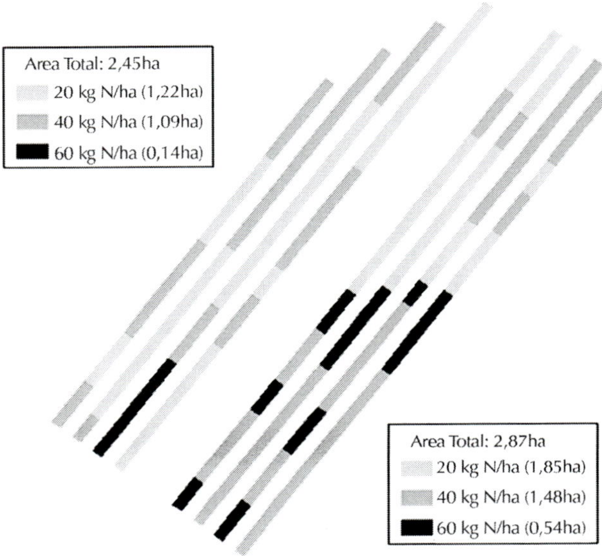

Figure 15: Sketch of the nitrogen application.

The harvest of the strips was realized using a combine John Deere 9650 STS with a 30 feet header and equipped with a yield monitor AgLeader PF3000 Advantage, what made possible to generate the yield map. The Figure 16 shows the yield map that was similar to the NDVI map. We can see in the map that each strip has a pair, because one strip is the 120 kg/ha of N and the other is the variable rate application. The yield varied from 1,500 to 5,500 kg/ha of wheat. But even the strips with 120 kg/ha of N had some parts of the field with lower yields, which means that these regions of low yield have some other problem and not nitrogen deficiency. Comparing with the application sketch the region of higher rates are same regions of high yields.

The left side of the experiment (first 4 pair of strips) had an average yield of 3,053 kg/ha using variable rate application, with 69.3% of saving in nitrogen, and 3,026 kg/ha using 120 kg/ha of N. The right side of the experiment had an average yield of 3,568 kg/ha using variable rate, with 42.5% of saving in nitrogen, and 3,546 kg/ha using 120 kg/ha of N. There was no statistical difference between the yield of fixed rate and variable rate, but the savings were high. The saving of nitrogen was higher in the region with lower yield potential.

As high is the yield goal just with the nitrogen supplied by the soil, with no additional N (low RI), in general, lower will be the N rates to reach maximum yields. If the RI for a field is low (RI < 1.1), that means the places with no N applied are similar with the places with N, and probably the response to an additional application will also be low. But if the RI is high (RI > 1.1), probably the crop will respond to the fertilization, so complementary rates should be applied [36].

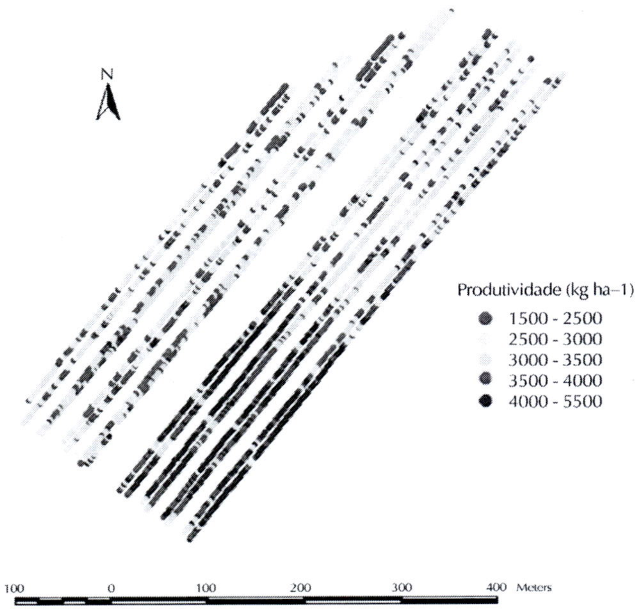

Figure 16: Yield map collected by the John Deere combine.

Ps: In each pair of strips, the left one is the 120 kg/ha of N and the right one is the variable rate application.

There is NDVI spatial variability even in the regions where fixed rates of nitrogen were applied, showing that crops respond in different ways inside the same field. The methodology used to apply nitrogen in variable rate, using the crop as an indicator and optical sensors to measure is really promising, being able to save nitrogen in places where there is less use by the plants.

However, optical sensors do not consider yield potential. Reflected light by the crop in a high yield zone can show an appropriate content of N at the moment of measurement, but N can still be deficient before maturation. In a low yielding zone, reflected light may suggest a higher need for N, inducing an excess rate application for plants that are limited by other factors [38].

Large differences were found between two corn fields, one with yields <12,000 kg/ha and the other with yields >17,000 kg/ha [39]. These authors suggested that considering climate conditions between planting date and the measurements could lead to a better relationship between yield estimation and actual yield because climate varies among areas and among years. Even when data from different areas and years are combined, the resulting model may not be reliable due spatial and temporal differences. These differences are some of the primary problems in developing a model that can be used widely to estimate yield by optical sensors. They also highlight that there is much room for improvement.

Variables that affect N availability were included, such as organic matter and soil exchange capacity, and variables that are related to N demand by the crop, such as solar radiation [40]. Additionally, the variability of rain distribution and water stored in soil at the time of planting can affect yield and management decisions [41].

Based on the observations of long-term experiments, it was found that the treatments with higher yield did not always correspond to higher N application rates. In some years, maximum yields were observed with low N application rates. In 3 long-terms experiment, there was no relationship between RI and yield. These results indicated that RI and yield are independent, and so they need to be estimated separately. Any practice that does not consider the independence of RI and yield can lead to misinformed recommendations. If yield potential varies among years along with N demand, the obvious solution to the increased efficiency of N recommendations is to be able to estimate crop response to N and yield potential [42].

CONCLUSIONS

The NDVI obtained from optical sensors are data that look very promising to map and analyze spatial variability of crop's development and for nitrogen management. It is possible to find significant regressions between NDVI and many crops' parameters. The technology of using optical sensors to recommend nitrogen showed high savings of the nutrient, making agriculture more profitable and with lower risk of water contamination by nitrate. Research with the application of optical sensors in agriculture is still scarce and need to expand to a wide range of conditions, for many other crops and applications.

REFERENCES

1. Liu WTH. Remote Sensing Application (in Portuguese). Campo Grande: Uniderp; 2007.
2. Tumbo SD, Wagner DG, Heinemann PH. Hyperspectral characteristics of corn plants under different chlorophyll levels. Transactions of the ASAE 2002;45(3) 815-823.
3. Blackmer TM, Schepers JS, Varvel GE, Walter-Shea EA. Nitrogen deficiency detection using reflected shortwave radiation from irrigated corn canopies. Agronomy Journal 1996;88(1) 1-5.
4. Baret F, Guyot G. Potentials and limits of vegetation indices for LAI and APAR assessment. Remote Sensing of Environment 1991;35(1) 161-173.
5. Boissard P, Pointel JG, Tranchefort J. Estimation of the ground cover ratio of a wheat canopy using radiometry. International Journal of Remote Sensing 1992;13(9) 1681-1692.
6. Tucker CJ, Holben BN, Elgin JH Jr., McMurtrey, J.E. Remote sensing of total dry-matter accumulation in winter wheat. Remote Sensing of Environment 1981;11(1) 171-189.

7. Waheed T, Bonnell RB, Prascher SO, Paulet E. Measuring performance in precision agriculture: CART – A decision tree approach. Agricultural Water Management 2006;84(1) 173-185.
8. Fischer RA, Howe GN, Ibrahim Z. Irrigated spring wheat and timing and amount of nitrogen fertilizer: I - Grain yield and protein content. Field Crops Research 1993;33(1) 37-56.
9. Solie JB, Stone ML, Raun WR, Johnson GV, Freeman K, Mullen R, Needham DE, Reed S, Washmon CN. Real-time sensing and N fertilization with a field scale GreenSeeker applicator: Proceedings of the International Conference on Precision Agriculture, Minneapolis, 2002.
10. Moreira RC. Influência do posicionamento e da largura de bandas de sensores remotos e dos efeitos atmosféricos na determinação de índices de vegetação. Master's dissertation. Instituto Nacional de Pesquisas Espaciais; 2000.
11. Jordan CF. Derivation of leaf area index from quality of light on the forest floor. Ecology 1969;50(4) 663-666.
12. Rouse JW, Haas RH, Schell JA, Deering DW. Monitoring vegetation systems in the great plains with ERTS: Proceedings of ERTS Symposium, Washington, DC, 1973.
13. Woolley JT. Reflectance and transmittance of light by leaves. Plant Physiology 1971;47(5) 656–662.
14. Piekielek WP, Fox RH, Toth JD, Macneal KE. Use of a chlorophyll meter at the early dent stage of corn to evaluate nitrogen sufficiency. Agronomy Journal 1995;87(3) 403-408.
15. Wright DL, Rasmussen VP, Ramsey RD, Baker DJ., Ellsworth JW. Canopy reflectance estimation of wheat nitrogen content for grain protein management. GIScience and Remote Sensing 2004;41(4) 287-300.
16. Steven MD. Correcting the effects of field of view and varying illumination in spectral measurements of crops. Precision Agriculture 2004;5(1) 55-72.
17. Stone ML, Solie JB, Raun WR, Whitney RW, Taylor SL, Ringer JD. Use of spectral radiance for correcting in–season fertilizer nitrogen deficiencies in winter wheat. Transactions of the ASAE 1996;39(5) 1623–1631.

18. Taylor SL, Raun WR, Solie JB, Johnson GV, Stone ML, Whitney RW. Use of spectral radiance for correcting nitrogen deficiencies and estimating soil test variability in an established Bermuda grass pasture. Journal of Plant Nutrition 1998;21(11) 2287–2302.

19. Ayala-Silva T, Beyl CA. Changes in spectral reflectance of wheat leaves in response to specific macronutrient deficiency. Advances in Space Research 2005;35(2) 305-317.

20. Raun WR, Solie JB, Johnson GV, Stone ML, Lukina EV, Thomason WE, Schepers JS. In-season prediction of potential grain yield in winter wheat using canopy reflectance. Agronomy Journal 2001;93(1) 131-138.

21. Thai CN, Evans MD, Deng X, Theisen AF. Visible & NIR imaging of bush beans grown under different nitrogen treatments. ASAE 1998;Paper 98-3074.

22. Sui R, Wilkerson JB, Hart WE, Wilhelm LR, Howard DD. Multi-spectral sensor for detection of nitrogen status in cotton. Applied Engineering in Agriculture 2005;21(2) 167-172.

23. Min M, LEE WS. Determination of significant wavelengths and prediction on nitrogen content for citrus. Transactions of the ASAE 2005;48(2) 455-461.

24. Kim Y, Evans RG, Waddell J. Evaluation of in-field optical sensor for nitrogen assessment of barley in two irrigation systems. ASAE 2005;Paper, PNW05-1004.

25. Frasson FR, Molin JP, Povh FP, Salvi JV. Temporal behavior of NDVI measured with an active optical sensor for different varieties of sugarcane. Revista Brasileira de Engenharia de Biossistemas 2007;1(1) 237-244.

26. Stafford JV. An investigation into the within-field spatial variability of grain quality: Proceedings of the European Conference on Precision Agriculture, Odense, 1999.

27. Tisdale SL, Nelson WL, Beaton JD, Halvin JL. Soil fertility and fertilizers. New York: Macmillan; 1993.

28. Raij B van. Soil fertility and fertilization (in Portuguese). São Paulo: Agronômica Ceres; 1991.

29. Gastal F, Lemaire G. N uptake and distribution in crops: an agronomical and ecophysiological perspective. Journal of Experimental Botany 2002;53(370) 789-799.

30. Schächtl J, Huber G, Maidl FX, Sticksel E, Schulz E, Haschberger P. Laser-induced chlorophyll fluorescence measurements for detecting the nitrogen status of wheat (Triticum aestivum L.) canopies. Precision Agriculture 2005;6(1) 143-156.

31. Solari F. Developing a crop based strategy for on-the-go nitrogen management in irrigated cornfields. PhD thesis. University of Nebraska; 2006.

32. Fageria NK, Baligar VC. Enhancing nitrogen use efficiency in crop plants. Advances in Agronomy 2005;88(1) 97-185.

33. Johnson GV, Raun WR. Nitrogen response index as a guide to fertilizer management. Journal of Plant Nutrition 2003;26(2) 249-262.

34. Inman D, Khosla R, Westfall DG, Reich R. Nitrogen uptake across site specific management zones in irrigated corn production systems. Agronomy Journal 2005;97(1) 169-176.

35. Raun WR, Solie JB, Johnson GV, Stone ML, Mullen RW, Freeman KW, Thomason WE, Lukina EV. Improving nitrogen use efficiency in cereal grain production with optical sensing and variable rate application. Agronomy Journal 2002;94(4) 815-820.

36. Mullen RW, Freeman KW, Raun WR, Johnson JV, Stone ML, Solie JB. Identifying an In-Season Response Index and the Potential to Increase Wheat Yield with Nitrogen. Agronomy Journal 2003;95(2) 347-351.

37. Rice CW, Havlin JL. Integrating mineralizable nitrogen indices into fertilizer nitrogen recommendations. In: Havlin JL, Jacobsen JL. Soil testing: Prospects for improving nutrient recommendations. Madison: Soil Science Society of America; 1994. p.1-14.

38. Lowenberg-DeBouer J. The Management Time Economics of On-the-go Sensing for Nitrogen Application, SSMC Newsletter, May 2004, Purdue University: http://www.agriculture.purdue.edu/ssmc/Frames/SSMC_May_2004_newsletter.pdf (accessed 5 July 2013).

39. Inman D, Khosla R, Reich RM, Westfall DG. Active remote sensing and grain yield in irrigated maize. Precision Agriculture 2007;8(4-5) 241-252.

40. Liu Y, Swinton SM, Miller NR. In site-specific yield response consistent over time? Does it pay? American Journal of Agricultural Economics 2006;88(2) 471-483.

41. Moeller C, Asseng S, Berger J, Milroy SP. Plant available soil water at sowing in Mediterranean environments—Is it a useful criterion to aid nitrogen fertilizer and sowing decisions? Field Crop Research 2009;114(1) 127-136.

42. Raun WR, Solie JB, Stone ML. Independence of yield potential and crop nitrogen response. Precision Agriculture 2011;12(4) 508-518.

Chapter

10

Cost-Effective Emission Abatement in Europe Considering Interrelations in Agriculture

Corjan Brink[1,2], Ekko van Ierland[2], Leen Hordijk[3], and Carolien Kroeze[3]

[1]National Institute of Public Health and the Environment (RIVM), P.O. Box 1, 3720 BA Bilthoven, The Netherlands;

[2]Environmental Economics and Natural Resources Group, Wageningen University, Hollandseweg 1, 6706 KN Wageningen, The Netherlands;

[3]Environmental Systems Analysis Group, Wageningen University, Diedenweg 18, 6703 GW Wageningen, The Netherlands

ABSTRACT

Agriculture is an important source of ammonia (NH_3), which contributes to acidification and eutrophication, as well as emissions of the greenhouse gases nitrous oxide (N_2O) and methane (CH_4). Controlling emissions of one of these pollutants through application of technical measures might have an impact (either beneficial or adverse) on emissions of the others. These side effects are usually ignored in policy making.

This study analyses cost-effectiveness of measures to reduce acidification and eutrophication as well as agricultural emissions of N_2O and CH_4 in Europe, taking into account interrelations between abatement of NH_3, N_2O, and CH_4 in agriculture.

The model used is based on the Regional Air pollution Information and Simulation (RAINS) model for air pollution in Europe, which includes emissions, abatement options, and atmospheric source-receptor relationships for pollutants contributing to acidification and eutrophication. We used an optimisation model that is largely based on the RAINS model but that also includes emissions of N_2O and CH_4 from agriculture and technical measures to reduce these emissions. For abatement options for agricultural emissions we estimated side effects on other emissions. The model determines abatement strategies to meet restrictions on emission and/or deposition levels at the least cost.

Cost-effective strategies to reduce acidification and eutrophication in Europe were analysed. We found that NH_3 abatement may cause an increase in N_2O emissions. If total agricultural N_2O and CH_4 emissions in Europe were not allowed to increase, cost-effective allocation of emission reductions over countries in Europe changed considerably.

INTRODUCTION

Many European countries agreed to reduce emissions contributing to acidification, eutrophication, and ground-level ozone (Gothenburg Protocol 1]) and greenhouse gas emissions (Kyoto Protocol [2]). Interrelations exist between policies for reducing emissions of air pollutants and greenhouse gases [3]. Several recent studies focus on ancillary benefits of efforts to reduce CO_2 emissions from fossil fuel use

[4]. In earlier work, we showed the existence of interrelations between acidification and greenhouse gas mitigation policies in European agriculture [5, 6, 7].

For greenhouse gases, the location of emissions is not important for their environmental impact. For pollutants contributing to acidification and eutrophication, however, the environmental impact depends on emission location and atmospheric transport before deposition. Moreover, the same level of deposition may have different environmental impacts at different locations because of differences in ecosystem sensitivities. A consideration of the side effects of air pollutants on emissions may therefore change the cost-effective geographical distribution of greenhouse gas emission reductions [8].

In this study we extend earlier work on interrelations between abatement of ammonia (NH_3), nitrous oxide (N_2O), and methane (CH_4) from European agriculture [5, 6, 7]. Here we also consider emissions of acidifying compounds other than NH_3 (namely sulphur dioxide [SO_2] and nitrogen oxides [NO_x]), the location of emissions and environmental effects and atmospheric transport of emissions in between, and environmental sensitivity of various locations. We analyse the impact of interrelations among abatement policies in agriculture on the cost-effective allocation of abatement over sources of emissions.

Model

A comparative static optimisation model was used to determine cost-effective abatement strategies in Europe for acidification, eutrophication, and agricultural emissions of N_2O and CH_4 simultaneously, considering interrelations between abatement activities in the agricultural sector. The model is largely based on the RAINS (Regional Air pollution INformation and Simulation) model developed at the International Institute for Applied Systems Analysis as a tool for integrated assessment of alternative strategies to reduce air pollution in Europe [9, 10]. The RAINS model includes data to estimate emissions of SO_2, NO_x, and NH_3 in Europe, as well as a large number of technical measures to abate these emissions. The model we used for this analysis is described in detail in Brink et al. [11]. This section will briefly describe this basic model and present some extensions to include several environmental effects and atmospheric transport of pollutants from the location of emissions to the location of the environmental impact.

Basic Model

The model is used to calculate emissions of various pollutants and determine cost-effective abatement strategies for restrictions on emission levels. Unabated emissions are calculated from (exogenous) data on various economic activities and can be reduced by applying abatement options at various sources. The model minimises total cost of abatement:

$$\min \sum_{k \in K} C_k(v_k) \tag{1}$$

where vk is a vector with elements ak,n (n ∈ N);

$$C_k(v_k) = \sum_{n \in N} a_{k,n}.\overline{X}_k \ \forall k \in K \tag{2}$$

Subject to

$$\sum_{k \in K} e_{k,p} \leq \overline{E}_p \ \forall p \in P, \tag{3}$$

$$e_{k,p} = e_{k,p}.\overline{X}_k.\left(1 - \sum_{n \in N} a_{k,n}.\rho_{k,n,p}\right) \forall p \in P, k \in K, \tag{4}$$

$$\sum_{n \in N} a_{k,n} \leq 1 \ \ \forall k \in K, \text{and} \tag{5}$$

$$\sum_{n \in N} a_{k,n} \geq 0 \ \ \forall k \in K,\ n \in N. \tag{6}$$

Decision variables in the model are application rates ($a_{k,n}$) for abatement options (n) at source k (lower case letters indicate individual elements of sets and upper case letters indicate sets). Abatement costs at source k (C_k) depend on the (exogenous) activity level ($\overline{X}_k$) and on application rates ($a_{k,n}$) and per unit costs ($\gamma_{k,n}$) for all abatement options ($n \in N$) (Eq. 2). Annual emissions of pollutant p from source k (ek,p) depend on activity level ($\overline{X}_k$), emission factor ($\varepsilon_{k,p}$), application rate ($a_{k,n}$), and effectiveness ($\rho_{k,n,p}$) of abatement options (Eq. 4). An upper level for emissions of each pollutant p is indicated by E – p (Eq. 3) and application rates ($a_{k,n}$) are constrained in Eq. 5 and 6. Emissions can be reduced by applying add-on abatement techniques. For each source, the model chooses application rates for all abatement options included in set N such that emission targets are obtained at minimum cost. Abatement options are not attributed to a specific pollutant; instead, an option may have impacts on various pollutants, either reducing or increasing emissions, but reducing emissions of at least one pollutant. The effect of abatement option n on emissions of pollutant p from source k is given as the fraction of unabated emissions that is reduced ($\rho_{k,n,p}$). Emissions of pollutant p can be reduced ($0 < \rho_{k,n,p} \leq 1$; i.e., a reduction to a maximum of 100%), unaffected ($\rho_{k,n,p} = 0$), or increased ($\rho_{k,n,p} < 0$; i.e., an increase can be more than 100%). The changes in emissions by abatement option n at source k depend linearly on the level of unabated emissions and on the application rate ($\varepsilon_{k,p} \times \overline{X} \times a_{k,n} \times \rho_{k,n,p}$).

There may be interaction between different measures to reduce emissions. For instance, measures may exclude, reinforce, or weaken each other, or some measures can only be applied in combination with others. To take this into consideration, possible combinations of measures and their combined effect were included as new abatement options. This may result in a large number of abatement options included in set N. Moreover, a certain level of detail about sources and abatement options is required in the model because side effects of emission-reduction measures may depend on specific characteristics of sources and measures. As a result, the model includes a large number of variables. To be able to solve such a large problem, the model applies a (stepwise) linear approach to calculate emissions, costs, and effects of abatement.

In the optimisation procedure, the abatement options that are applied for specific emission constraints are determined. Cost-effectiveness of abatement strategies depends on the cost of abatement options, the magnitude of effects of abatement options on emissions of various pollutants, and reduction targets for emissions of various pollutants.

Multiple Environmental Targets

To reflect the contribution of various pollutants to different types of environmental damage, the model includes indicators for environmental problems ($q \in Q$), reflecting the size of the environmental impact. These indicators depend linearly on the emissions of pollutants.* The indicator for environmental problem q (h_q) is determined by:

$$h_q = \sum_{k \in K} \sum_{p \in P} \psi_{p,q} . e_{k,p} \tag{7}$$

where $\psi_{p,q}$ represents the contribution of one unit of pollutant p to environmental problem q in terms of the indicator used for this environmental problem (for example, $\psi_{p,q}$ may be the global warming potential (GWP) for greenhouse gas p, indicating the relative contribution of this pollutant to global warming). Constraints in the model can be specified in terms of these indicators:

$$h_q \leq \bar{H}_q \quad \forall q \in Q \tag{8}$$

where $\bar{H}_q$ indicates the target level for indicator h_q.

Location of Emissions and Effects

Emissions occur at different locations. For some pollutants (e.g., air pollutants) the environmental impact depends on the location of the emissions and transport of pollutants through the atmosphere, whereas for others (e.g., greenhouse gases) the location does not matter because emissions are uniformly mixing in the atmosphere. In the case of air pollutants, the environmental impact does not necessarily occur at the place of emissions because pollutants may be transported through the atmosphere over long distances before they have an effect on environmental quality [9].

For pollutants of this type, it is important to take into account the location of emissions, the atmospheric transport, and the location of the environmental effect when determining cost-effective abatement strategies. To this end, the model is extended to include a set I of source locations (i.e., where the emissions occur) and a set J of receptor locations (i.e., where the environmental effects occur).** All sources (and hence emissions) are assigned to a source location ($i \in I$). Consequently, the dimension i is added to economic activity level ($\overline{X}_k$), emission level ($e_{k,p}$), and application rate ($a_{k,n}$) (i.e., they become $\overline{X}_{i,k}$, $e_{i,k,p}$, and $a_{i,k,n}$, respectively).

Foot Note

*This is an accepted approach for some environmental problems, such as acidification, and for adding up emissions of various greenhouse gases. For other environmental problems, such as the formation of ground-level ozone, which is the result of a nonlinear relationship between several pollutants, this approach is not possible. In principle, this kind of nonlinear relationship can be included in the model using a stepwise linear approach.

**Both source [i] and receptor [j] locations can be countries, but a higher or lower level of aggregation is possible as well. Moreover, the level of aggregation is not necessarily the same for source and receptor.

Because of different characteristics of sources at different locations, both the emission factors and the parameters for the cost and effect of abatement options are also location specific (i.e., $\varepsilon_{k,p}$, $\gamma_{k,n}$, and $\rho_{k,n,p}$ are replaced by $\varepsilon_{i,k,p}$, $\gamma_{i,k,n}$, and $\rho_{i,k,n,p}$, respectively). In accordance with these changes, Eq. 1, 2, 4, 5, and 6 are respectively replaced by

$$\min \sum_{i \in I} \sum_{k \in K} C_{i,k}\left(v_{i,k}\right) \tag{9}$$

where $v_{i,k}$ is a vector with elements $a_{i,k,n}$ ($n \in N$);

$$C_{i,k}\left(v_{i,k}\right)=\sum_{n\in N} a_{i,k,n}.\gamma_{i,k,n}.\overline{X}_{i,k}; \tag{10}$$

$$e_{i,k},p=\varepsilon_{i,k,p}.\overline{X}_{i,k}.\left(1-\sum_{n\in N} a_{i,k,n}.\rho_{i,k,n,p}\right); \tag{11}$$

$$\sum_{n\in N} a_{i,k,n}\le 1\ \forall\ i\in I, k\in K;\text{and} \tag{12}$$

$$a_{i,k,n}\ge 0\ \forall i\in I,\ k\in K,\ n\in N. \tag{13}$$

The process of atmospheric transport of pollutants is included in the model, assuming a linear relationship between emissions of pollutant p at source location i that are deposited at receptor location j. The constant parameter $\tau_{i,j,p}$ represents the fraction of emissions of pollutant p at i that is deposited at j. The same approach is used in the RAINS model to include atmospheric transport of air pollutants [9, 10]. Restrictions can be specified for emissions and environmental effects at various (source or receptor) locations or for overall emissions and environmental effects.

COST-EFFECTIVE ABATEMENT IN EUROPE

The model was used to analyse cost-effective policies to reduce acidification and eutrophication in Europe, as well as emissions of greenhouse gases such as N_2O and CH_4 from agriculture, considering interrelations between technical abatement measures in the agricultural sector. Acidification in Europe is mainly the result of emissions of SO_2, NO_x, and NH_3. Both NO_x and NH_3 also play an important role in eutrophication. SO_2 and NO_x emissions are to a large extent caused by fossil fuel combustion, whereas the major source of NH_3 is agriculture. Important gases contributing to global warming are carbon dioxide (CO_2), N_2O, and CH_4.

In this study only greenhouse gas emissions related to agricultural activities are included. In Europe, N_2O and CH_4 are to a large extent the result of agricultural activities. In the 1990s, agriculture was responsible for about 8% of total greenhouse gas emissions in Europe [12]. European agriculture is a minor source of CO_2 emissions, but may play a role in reducing CO_2 emissions by replacing fossil fuels with biomass and by removing CO_2 through sinks [13]; however, these considerations are not included in this analysis.

Application of the Model

Environmental problems included in the model (set Q) are acidification, eutrophication, and greenhouse gas emissions; only emissions of N_2O and CH_4 from agriculture are considered. Pollutants considered (set P) are SO_2, NO_x, NH_3, N_2O, and CH_4. For acidification and eutrophication, the analysis considers atmospheric transport of pollutants before they are deposited. Moreover, targets for acidification and eutrophication take into account ecosystem sensitivity at the location of deposition. For N_2O and CH_4 from agriculture, restrictions are on the sum of CO_2-equivalent emissions of these gases,* which is justified because N_2O and CH_4 are uniformly mixed in the atmosphere. Indicators for the environmental impacts (h_q) are acid equivalents (acidification), nitrogen (eutrophication), and CO_2-equivalents (greenhouse gas emissions).

Furthermore, the model includes emissions from 36 European countries (set I). For SO_2, NO_x, and NH_3, the model considers atmospheric transport of emissions from each country to about 500 grids of 150 × 150 km in Europe (set J). The source-receptor relationship ($\tau_{i,j,p}$, indicating the fraction of pollutant p emitted in country i that is deposited in grid j) was taken from the RAINS model. Source-receptor matrices in the RAINS model were derived from a model of long-range transport of air pollution developed by EMEP (the Cooperative Programme for Monitoring and Evaluation of the Long-Range Transmission of Air Pollutants in Europe) [10, 14].

Within each country, the model distinguishes various sources of emissions (set K). The focus of this study is on the agricultural sector;

therefore, agricultural activities (as sources of NH_3, N_2O, and CH_4) are considered in more detail than nonagricultural activities (as sources of SO_2 and NO_x). For agriculture, the model includes 14 sources of emissions in each country (viz. 11 animal categories, 2 types of fertiliser used, and production of nitrogen fertilisers). For SO_2 and NO_x, all sources in a country were aggregated in one source for SO_2 and one source for NO_x emissions (with activity levels $\overline{X}_{i,k}$ for these sources representing emissions of SO_2 and NO_x). Activity levels for the various sources in each country ($\overline{X}_{i,k}$) and the associated emission factors for NH3 ($\varepsilon_{i,k}$, NH_3) were taken from databases in the RAINS model. This model includes data for 1990, 1995, and projections up to 2010.

Agricultural emissions of N_2O and CH_4 were estimated on the basis of the databases in the RAINS model using emission factors as described in Brink et al. [5].

The model includes a set N of abatement options. Information on costs per unit of activity ($\gamma_{i,k,n}$) and effects on emissions ($\rho_{i,k,n,p}$) was taken from the RAINS model[10, 15] and from studies for the EU[16,17]. For abatement options applicable to agricultural activities, we estimated effects on other emissions [5, 7].

For each grid cell, we used the fifth percentiles of the critical loads of sulphur and nitrogen (both acidifying and nutrient). **These values were calculated using critical loads for many ecosystems that occur within a grid cell [19, 20]. If the deposition in a grid cell does not exceed these values, 95% of the ecosystem area within the grid cell is protected.

Emissions and abatement costs in 2010 were estimated for four scenarios. First, the no-control (NOC) scenario assumes no technical abatement measures applied in any European country (this scenario also excludes measures that have to be taken in the future according to adopted national and international legislation for emission control). Second, the 40% gap-closure (GAP) scenario reflects a reduction in emissions of acidifying compounds in Europe such that for each grid cell, the critical load exceedance*** for acidification and eutrophication in the NOC scenario is reduced by at least 40% at minimum total abatement cost. Third, the 40% gap-closure-plus (GAP^+) scenario requires a 5% reduction in the sum of N_2O and CH_4 emissions (in CO_2-equivalents) from agriculture in Europe relative to their NOC levels by measures additional to the abatement options applied in the cost-effective solution in the GAP scenario. Finally, the 40% gap-closure double-plus (GAP^{++}) scenario combines the 40% gap-closure requirement for acidification and eutrophication with the 5% reduction requirement for N_2O and CH_4 emissions (in CO_2-equivalents) from agriculture in Europe. In the GAP^{++} scenario, cost-effective abatement strategies are determined for both the 40% gap-closure and the 5% emission-reduction requirements simultaneously, considering interrelations and side effects of abatement options in agriculture.

Foot Note

*N_2O and CH_4 emissions are converted to CO_2-equivalent emissions using the Intergovernmental Panel on Climate Change's GWPs for a time horizon of 100 years [310 for N_2O; 21 for CH_4].

**A critical load for acidification and eutrophication has been defined as "a quantitative estimate of an exposure to one or more pollutants below which significant harmful effects on specified sensitive elements of the environment do not occur according to present knowledge."[18, 19].

***A critical load exceedance is defined as "the sum of the nitrogen and sulphur deposition reduction required to reach the critical load function by the 'shortest' path."[19].

RESULTS AND DISCUSSION

To achieve the reduction in critical-load exceedance, emissions of SO_2, NO^x, and NH_3 have to be reduced considerably (Table1).

NH_3 abatement without restrictions on N_2O and CH_4 emissions (GAP) caused an increase in total N_2O emissions from European agriculture of 8% relative to NOC (Table 1) as a result of negative side effects of NH_3 abatement options. Total agricultural emissions of CH_4 decrease a little as a result of the side effects of NH_3 abatement. The net increase in N_2O and CH_4 emissions from agriculture in CO_2-equivalents is 4%.

Adding a reduction target for N_2O and CH_4 emissions from European agriculture (GAP^+ and GAP^{++}) obviously results in higher total abatement costs than in the GAP scenario (Table 1).

Table 1: Estimated Emissions and Total Abatement Cost in Europe in 2010 for Four Scenarios (NOC, GAP, GAP^+, and GAP^{++})

	NOC		GAP		GAP^+		GAP^{++}	
SO_2	15.9	(100%)	12.4	(78%)	12.4	(78%)	12.4	(78%)
NO_x	16.7	(100%)	13.6	(82%)	13.4	(80%)	12.9	(78%)
NH_3	6.7	(100%)	4.8	(72%)	4.9	(73%)	4.9	(74%)
N_20^a	1.7	(100%)	1.8	(108%)	1.6	(95%)	1.6	(95%)
$CH_4{}^a$	16.6	(100%)	16.3	(98%)	15.7	(95%)	15.6	(94%)
Abatement cost[b]			41.7		55.8		47.6	

Note: Emissions are in million tons SO_2, NO_2, NH_3, N_2O, and CH_4/year; abatement costs are in billion Euros1990/year; percentages given represent emissions relative to emissions in NOC scenario.

[a]Emissions from agricultural activities only.

[b]Total cost for technical measures to reduce emissions in Europe

Results for GAP^+ are different from GAP^{++} because GAP^+ analyses optimal abatement strategies for acidification and greenhouse gases separately; GAP^{++} reflects an integrated approach reducing acidifying emissions and agricultural N_2O and CH_4 emissions simultaneously, considering interrelations between abatement of NH_3, N_2O, and CH_4 in European agriculture. In GAP^+, the GAP abatement strategy is maintained, including abatement options that cause an increase in N_2O emissions. The required reduction in N_2O and CH_4 can only be obtained by applying additional abatement options so that the reduction target for N_2O and CH_4 is obtained without violating restrictions for acidification and eutrophication. In GAP^{++}, total abatement costs are 15% lower than in GAP^{+} because in GAP^{++}, it is possible to apply abatement options that simultaneously reduce NH_3 and N_2O emissions instead of abatement options that reduce NH_3 emissions but cause a simultaneous increase in N_2O emissions. The costeffective allocation of abatement over countries also differs between GAP^{+} and GAP^{++} (Table 2). In particular, the change in abatement costs in GAP^{++} relative to GAP^{+} differs between countries.

For most countries we calculated lower total abatement costs in GAP^{++} than in GAP^{+}, but in some countries (e.g., Belgium and Finland) calculated costs are higher in GAP^{++} than in GAP^{+} (Table 2).

Table 2: Differences in Country-Specific Emissions and Abatement Costs between GAP++ and GAP+ Scenarios[a,b]

	NO, (GAP•-Git•)		NH, (GAP- -GAP′)		N,O′ (GAP•-GAF•)		(GAP- -GAP′)		Abatement Cost′ (GAP--GAP•)	
Albania	0	(1%)	0	(0%)	-0	(0%)	-1	(-1%)	0	(2%)
Austria	—		0	(0%)	-1	(-5%)	4	(2%)	-19	(-4%)
Belams	-10	(-6%)	-3	(-3%)	2	(6%)	-3	(-1%)	-290	(-30%)
Belgium	-8	(-6%)	-2	(-3%)	-1	(-5%)	-5	(-2%)	186	(8%)
Bosnia Herzegovina	-8	(-16%)	0	(0%)	-0	(-1%)	0	(0%)	9	(12%)
Bulgaria	-1	(0%)	1	(1%)	1	(4%)	-2	(-2%)	-296	(-75%)
Croatia	-9	(-17%)	1	(4%)	0	(2%)	0	(0%)	-65	(-27%)
Czech Republic	—		1	(1%)	1	(5%)	-6	(-2%)	-278	(-15%)
Denmark	—		-5	(-8%)	0	(1%)	-11	(-6%)	-82	(-22%)
Estonia	8	(12%)	0	(0%)	-0	(0%)	-1	(-3%)	-2	(-11%)
Finland	—		-1	(-3%)	-0	(-1%)	-1	(-2%)	5	(21%)
France	-7	(-1%)	-12	(-2%)	3	(2%)	15	(1%)	-830	(-21%)
Germany	—		-5	(-1%)	-5	(-4%)	13	(1%)	-25	(-1%)
Greece	—		1	(1%)	0	(0%)	-1	(-1%)	-28	(-59%)
Hungary	—		4	(6%)	-2	(-5%)	1	(1%)	-327	(-18%)
Ireland	—		-0	(0%)	1	(2%)	-1	(0%)	-223	(-41%)

Italy	-0	(0%)	6	(2%)	-9	(-10%)	-14	(-2%)	-113	(-1%)
Latvia	—		-0	(-1%)	-0	(-1%)	-1	(-2%)	2	(6%)
Lithuania	-8	(-8%)	0	(1%)	1	(4%)	-2	(-1%)	-146	(-54%)
Luxembourg	—		0	(1%)	0	(3%)	0	(0%)	-11	(-7%)
Netherlands	—		-8	(-7%)	0	(0%)	4	(1%)	38	(4%)
Norway	13	(8%)	-1	(-3%)	0	(1%)	-0	(0%)	-164	(-48%)
Poland	-7	(-1%)	4	(1%)	3	(3%)	-12	(-1%)	-579	(-9%)
Portugal	—		-0	(0%)	0	(3%)	-1	(-1%)	-91	(en)
Republic of	—		0	(0%)	-0	(0%)	-0	(0%)	-1	(-26%)
Macedonia										
Republic of Moldova	-2	(-4%)	0	(1%)	0	(4%)	-0	(0%)	-125	(-67%)
Romania	-35	(-14%)	4	(2%)	-2	(-2%)	-3	(0%)	-338	(-21%)
Russia	-189	(-8%)	43	(6%)	1	(0%)	-18	(-1%)	-1085	(-43%)
Slovak Republic	-41	(41%)	3	(10%)	0	(2%)	-3	(-3%)	327	(122%)
Slovenia	—		1	(5%)	0	(3%)	0	(0%)	-61	(-43%)
Spain	—		-0	(0%)	2	(2%)	—		-541	(-57%)
Sweden	—		0	(0%)	1	(5%)	-4	(-3%)	-124	(-59%)
Switzerland	—		-0	(0%)	0	(0%)	1	(0%)	-6	(-1%)
Ukraine	-159	(-16%)	25	(5%)	4	(3%)	-26	(-2%)	-1176	(-40%)
United Kingdom	56	(5%)	-7	(-3%)	5	(4%)	-14	(-1%)	-1678	(-22%)
YugosliMa	-21	(-23%)	4	(6%)	0	(1%)	-0	(0%)	-79	(-19%)
Total	-426	l-3%)	54	(1%)	6	(0%)	-93	(-1%)	-8215	(-15%)

Note: Emissions are given as kilotons NO_2, NH_3, N_2O, and CH_4 / year; abatement costs are in million Euro/year; percentages represent a percentage of GAP+ values.

[a]SO2 emissions calculated for the GAP+ and GAP++ scenarios are the same for all countries.

[b] — indicates that there is no change; 0 indicates an increase smaller than 0.5; –0 indicates a reduction smaller than 0.5.

[c]Emissions from agricultural activities only.

[d]Total cost for technical measures to reduce emissions in Europe.

CONCLUSIONS

Interrelations between policies aiming at different environmental problems may have an effect on cost-effective strategies to meet reduction targets for these environmental problems. This paper describes a stepwise linear, comparative, static optimization model that can be used to analyse cost-effective emission reduction strategies to meet various environmental targets, considering several pollutants, atmospheric transport of pollutants, a large number of abatement options, and their effects on emissions of several pollutants.

The model was used to analyse cost-effective strategies to reduce acidification and eutrophication in Europe on the basis of information from the RAINS model. Moreover, we analysed the impact of NH_3 abatement on emissions of greenhouse gases such as N_2O and CH_4 from agricultural activities. For abatement options applicable to agricultural activities, we estimated possible side effects on emissions. Reduction targets were specified for nitrogen and sulphur deposition exceeding critical loads for acidification and eutrophication in grid cells of 150 ↔ 150 km over Europe. We calculated a cost-effective allocation of abatement over sources to meet these targets with and without a restriction on emissions of N_2O and CH_4. Without a restriction on N_2O and CH_4 emissions, N_2O emissions were calculated to increase 8% as a result of NH_3 abatement. A 5% reduction target for agricultural N_2O and CH_4 emissions in Europe in addition to the acidification and eutrophication targets resulted in higher total abatement costs. We analysed a separate approach and an integrated approach to achieve the reduction targets for acidification and eutrophication and the reduction target for N_2O and CH_4 emissions. The separate approach first determines a costeffective abatement strategy for acidification and eutrophication targets only; this approach is maintained when subsequently an additional abatement strategy is determined to achieve the required reduction in N_2O and CH_4 emissions. The integrated approach determines a cost-effective abatement strategy for all targets simultaneously. NH_3 abatement options that involve an increase in N_2O emissions are maintained in the separate approach.

In the integrated approach, however, it is possible to apply NH_3 abatement options without an effect on N_2O or abatement options that simultaneously reduce N_2O and NH_3 emissions instead of NH_3 abatement options causing an increase in N_2O. Consequently, total abatement costs are 15% lower in the integrated approach than in the separate approach. Moreover, the allocation of abatement costs over countries is different, with most countries facing lower costs with the integrated approach than with the separate approach, but with some countries facing higher costs.

This study focuses on interrelations between acidification and greenhouse gas mitigation policies in the agricultural sector, considering technical emission-control options. Interrelations between these policies may, however, also exist in other sectors, and other areas of policy-making may be interrelated with these policies. Moreover, in addition to the technical control options, structural changes may play an important role in reducing greenhouse gas emissions in the longer term. Several of these aspects can be included in the model, although for some of these extensions the linearised approach might be unsuitable. Major limitations are data availability (in particular data on side effects of abatement technologies) and also the size of the problem (because of the large number of abatement options that have to be included). The case presented in this study indicates that interrelations between different environmental problems affect costeffective policies.

ACKNOWLEDGEMENTS

This research was funded by the Netherlands Organisation for Scientific Research. We thank Dr. Jean-Paul Hettelingh for providing us with critical loads data for Europe. We thank two anonymous reviewers for their useful comments on an earlier version of the manuscript.

REFERENCES

1. UNECE. (1999) Protocol to the 1979 Convention on Long-Range Transboundary Air Pollution to Abate Acidification, Eutrophication and Ground-Level Ozone. United Nations Economic Commission for Europe, Geneva, Switzerland.
2. UNFCCC. (1997) Kyoto Protocol to the United Nations Framework Convention on Climate Change. United Nations Framework Convention on Climate Change, Bonn, Germany.
3. Grennfelt, P., Hov, Ø., and Derwent, R.G. (1993) Second Generation Abatement Strategies for NOx, NH_3, SO_2, and VOC. IVL Report, B-1098. Swedish Environmental Research Institute (IVL), Stockholm.
4. Davis, D.L., Krupnick, A.J., and McGlynn, G. (2000) Ancillary benefits and costs of greenhouse gas mitigation: an overview. In Ancillary Benefits and Costs of Greenhouse Gas Mitigation: Proceedings of an IPCC Co-sponsored Workshop, 27–29 March 2000, Washington, D.C. Davis, D.L., Krupnick, A.J., and McGlynn, G., Eds. Organization for Economic Cooperation and Development, Paris. pp. 9–49.
5. Brink, C., Kroeze, C., and Klimont, Z. (2001) Ammonia abatement and its impact on emissions of nitrous oxide and methane. I. Method. Atmos. Environ., in press.
6. Brink, C., Kroeze, C., and Klimont, Z. (2001) Ammonia abatement and its impact on emissions of nitrous oxide and methane. II. Application for Europe. Atmos. Environ., in press.
7. Brink, C., Hordijk, L., Van Ierland, E.C., and Kroeze, C. (2000) Cost-effective N_2O, CH_4, and NH_3 abatement in European agriculture: interrelations between global warming and acidification policies. Paper presented at Ancillary Benefits and Costs of Greenhouse Gas Mitigation. Proc. IPCC Co-sponsored Workshop, 27–29 March 2000, Washington, D.C. Available at http://www.oecd.org/env/docs/cc/abworkshop-brink.pdf.
8. Pearce, D.W. (2000) Policy frameworks for the ancillary benefits of climate change policies. In Ancillary Benefits and Costs of Greenhouse Gas Mitigation: Proceedings of an IPCC Co-sponsored Workshop, 27–29 March 2000, Washington, D.C.

Davis, D.L., Krupnick, A.J., and McGlynn, G., Eds. Organization for Economic Cooperationand Development, Paris. pp. 517–560.

9. Alcamo, J., Shaw, R.W., and Hordijk, L. (1990) The RAINS Model of Acidification; Science and Strategies in Europe. Kluwer, Dordrecht, The Netherlands.
10. Amann, M., Bertok, I., Cofala, J., Gyarfas, F., Heyes, C., Klimont, Z., Makowski, M., Schöpp, W., and Syri, S. (1998) Cost-Effective Control of Acidification and Ground-Level Ozone. 6th Interim Report to the European Commission, DG-XI. International Institute for Applied Systems Analysis (IIASA), Laxenburg, Austria. http://www.iiasa.ac.at/~rains/.
11. Brink, C., Van Ierland, E.C., Hordijk, L., and Kroeze, C. (2001) Cost-Effectiveness of Interrelated Emission Reduction Strategies in Agriculture. Research Report. Environmental Economics and Natural Resources Group, Wageningen University, The Netherlands. Submitted to European Review of Agricultural Economics. Available at http://www.sls.wau.nl/enr/staff/brink/.
12. UNFCCC (2000) UNFCCC Greenhouse Gas Inventory Database. United Nations Framework Convention on Climate Change, Bonn, Germany. http://www.unfccc.de/.
13. Paustian, K., Cole, C.V., Sauerbeck, D., and Sampson, N. (1998) CO_2 mitigation by agriculture: an overview. Climatic Change 40, 135–162.
14. Barrett, K. and Berge, E. (1996) Transboundary Air Pollution in Europe. EMEP Status Report, January 1996, Oslo, Norway.
15. Klaassen, G. (1994) Options and costs of controlling ammonia emissions in Europe. Eur. Rev. Agric. Econ. 21, 219–240.
16. Hendriks, C.A., De Jager, D., and Blok, K. (1998) Emission reduction potential and costs for methane and nitrous oxide in the EU-15. Interim Report, M714. ECOFYS Utrecht, The Netherlands. http://www.ecofys.nl/reports/.
17. Bates, J. (2000) Economic evaluation of emission reductions of nitrous oxides and methane in agriculture in the EU: bottom-up analysis. AEA Technology Report, Culham, U.K. http:// europa.eu.int/comm/environment/.
18. Nilsson, J. and Grennfelt, P. (1988) Critical Loads for Sulphur and Nitrogen, NORD 1988:97, Copenhagen, Denmark.

19. Posch, M., De Smet, P.A.M., Hettelingh, J.-P., and Downing, R.J. (1999) Calculation and Mapping of Critical Thresholds in Europe: Status Report 1999. Report 259101009. Dutch National Institute of Public Health and Environmental Protection (Rijksinstituut voor Volksgezondheid and Milieu [RIVM]), Bilthoven, The Netherlands. http://www.rivm.nl/cce.
20. Hettelingh, J.-P., Posch, M., and De Smet, P.A.M. (2001) Multieffect critical loads used in multi-pollutant reduction agreements in Europe. Water Air Soil Pollut., in press.

Chapter 11

Fertilizer Nitrogen Recovery Efficiencies in Crop Production Systems of China with and without Consideration of the Residual Effect of Nitrogen

Xiaoyuan Yan[1], Chaopu Ti[1], Peter Vitousek[2], Deli Chen[3], Adrian Leip[4], Zucong Cai[5], and Zhaoliang Zhu[1]

[1]State Key Laboratory of Soil and Sustainable Agriculture, Institute of Soil Science, Chinese Academy of Sciences, Nanjing 210008, People's Republic of China

[2]Department of Biological Sciences, Stanford University, Stanford, CA 94305, USA

[3]Melbourne School of Land and Environment, The University of Melbourne, Melbourne, VIC 3010, Australia

[4]European Commission-Joint Research Centre, Institute for Environment and Sustainability, Climate Change Unit, Via E. Fermi 2749, I-21027 Ispra (VA), Italy

[5]School of Geography Sciences, Nanjing Normal University, Nanjing 210097, People's Republic of China

ABSTRACT

China is the world's largest consumer of synthetic nitrogen (N), where very low rates of fertilizer N recovery in crops have been reported, raising discussion around whether fertilizer N use can be significantly reduced without yield penalties. However, using recovery rates as indicator ignores a possible residual effect of fertilizer N—a factor often unknown at large scales. Such residual effect might store N in the soil increasing N availability for subsequent crops. The objectives of the present study were therefore to quantify the residual effect of fertilizer N in China and to obtain more realistic rates of the accumulative fertilizer N recovery efficiency (RE) in crop production systems of China. Long-term spatially-extensive data on crop production, fertilizer N and other N inputs to croplands in China were used to analyze the relationship between crop N uptake and fertilizer N input (or total N input), and to estimate the amount of residual fertilizer N. Measurement results of cropland soil N content in two time periods were obtained to compare the change in the soil N pool. At the provincial scale, it was found that there is a linear relationship between crop N uptake and fertilizer N input or total N input. With the increase in fertilizer N input, annual direct fertilizer N RE decreased and was indeed low (below 30% in recent years), while its residual effect increased continuously, to the point that 40–68% of applied fertilizer was used for crop production sooner or later. The residual effect was evidenced by a buildup of soil N and a large difference between nitrogen use efficiencies of long-term and short-term experiments.

INTRODUCTION

As an essential constituent of the building blocks of life, i.e., DNA, RNA and proteins, nitrogen (N) is required by all organisms and ecosystems. However, human activities have more than doubled the annual amount of reactive N entering terrestrial ecosystems since the preindustrial era (Galloway 1998, Smil 1999, Green *et al* 2004). This added N has contributed to soil acidification, eutrophication, global warming and stratospheric ozone depletion, as well as particulate matter and photo-oxidant formation (Vitousek *et al* 1997).

Synthetic N fertilizer is by far the largest source of anthropogenic reactive N worldwide (Fowler *et al* 2013); its applications made the doubling of world food production possible in the past four decades (Tilman 1999). However, much of the fertilizer N intended for crops is instead lost to the environment via denitrification, NH_3 volatilization, surface runoff and leaching (Raun and Johnson1999). The fraction of applied N actually utilized by crops is often low, and much of the remainder can be considered an expensive and environmentally damaging waste of N.

Nitrogen use efficiency (NUE) of crop production is often measured in term of the 'recovery efficiency' of applied N (RE_N), and is defined as (Cassman *et al* 2002):

$$(U_N - U_0) / F_N \qquad (1)$$

where U_N is the plant N uptake (kg ha^{-1}) measured in aboveground biomass at physiological maturity in a plot that received N at the rate of F_N (kg ha^{-1}), and U_0 is the N uptake measured in aboveground biomass in a plot without the addition of fertilizer N. By definition, RE_N is relatively easy to measure and thus often used as the indicator of NUE.

China is the largest user of fertilizer N in the world, and many field experiments have shown low RE_N of 30–35% in the 1990s (Zhu and Chen 2002) and 26–28% in 2001–05 for major cereal crops (Zhang *et al* 2007), relative to 52% in America and 68% in Europe (Ladha *et al* 2005). This low RE_N has stirred up wide discussion on whether fertilizer N use can be significantly reduced in China (Ju *et al* 2009, Liu *et al* 2011, Qiao *et al* 2012). Long-term experiments on the other hand often yield higher RE_N results (Jensen and Schjoerring 2011), suggesting that part of the N previously applied could be recovered. Even the long-term experiments understate the contribution of fertilizer N because their zero-N plots continue to receive environmental contributions (e.g., dry deposition) of fertilizer N from neighboring plots. In addition, RE_N is dependent on fertilization rate; hence, the RE_N of such long-term experiments may not represent the present day situation if the fertilization rate has increased significantly since the beginning of the experiment.

Although residual N can be traced using the ^{15}N-isotope technique, this method may incorrectly quantify the amount of residual fertilizer N because pool N substitution causes immobilization of ^{15}N fertilizer in soil and thus overestimates residual fertilizer N (Krupnik *et al* 2004). Therefore, many of the data available in the literature are of RE_N and only a few studies have produced data to quantify residual fertilizer N recovery in subsequent crops (Ladha *et al* 2005), and this is especially true for large-area estimates.

Soil nitrogen budgets (Leip *et al* 2011, Eurostat 2013) on the other hand require a more thorough quantification of N inputs (as organic fertilizers, atmospheric deposition, and biological N fixation), but allow the estimation of NUE without the need of a 'no-fertilizer' reference.

We hypothesized that there is a significant residual effect of fertilizer N in areas with continuously high N input, such as China. Thus, the objectives of the present study were, by using long-term spatially extensive N budget data, to: (1) evaluate the relationship between crop N uptake and applied fertilizer N; (2) quantify the long-term effect of residual fertilizer N; and (3) illustrate the temporal trend in both RE_N and the residual fertilizer N effect.

MATERIALS AND METHODS

Crop N Uptake and Synthetic N Input

Crop N uptake was estimated from the amount of crop harvests and residues and their N contents. The amount of crop residue was estimated from crop harvests by using residue/harvest ratios (supplementary table

S1, available atstacks.iop.org/ERL/9/095002/mmedia). The amounts of synthetic N fertilizer consumption and production of various crops at national and provincial scales were obtained from the China Agriculture Yearbook (see the supplementary material for references). Synthetic N fertilizer application rates, yields and cultivated area of major crops are shown in table S2. Data on arable land area of after year 1996 were obtained from Communique of Main Data on Land Use Survey (Ministry of Land and Resources of China *et al* 1999) and Communique on Land and Resources of China (Ministry of Land and Resources of China 2008, 2009). Arable land data for years 1980–96 were obtained from China Agriculture Yearbook and adjusted using data from Communique of Main Data on Land Use Survey (Ministry of Land and Resources of China *et al* 1999). See the supplementary material for details.

Non-synthetic N Input

Biological N_2 fixation was estimated on an area basis. Fixing rates were assumed at 80 kg N ha^{-1} yr^{-1} for soybean and peanut (Smil 1999) and 150 kg N ha^{-1} yr^{-1} for green manure (Yan *et al* 2003). Flooded rice fields provide a unique set of conditions for biological N_2 fixation by autotrophic N_2-fixing Cyanobacteria and heterotrophic N_2-fixing bacteria (Roger and Ladha 1992), and we used a fixation rate of 30 kg N ha^{-1} as default (Zhu and Wen 1992). The fixation rate for upland crops was assumed to be 15 kg N ha^{-1} (Burns and Hardy 1975).

China is one of the global heavy N-deposition areas, especially in the southeast of the country (Galloway *et al* 2008), however, there is no systematic nationwide monitoring network to derive geographical and temporal distributions of deposition rates. Previous studies suggested a relationship between N deposition and emission (Asman 1998, Goulding *et al* 1998). In the current study, based on the N emission database of Asia (Ohara *et al* 2007) along with published wet N deposition data for China, we developed and validated region-specific N emission/wet deposition relationships (Ti *et al* 2012). These relationships and the N emission inventory were then used to estimate wet N deposition rates of different provinces from 1980 to 2010.

Owing to difficulties with measurement, only a few data are available for dry deposition in China, and most of them indicate a dry: wet deposition ratio of about 2: 8 (Xie 2006, Lü and Tian 2007, Yang *et al* 2010). However, in these studies, not all dry deposition N species were accounted for, and thus they likely underestimated dry deposition. Considering that dry deposition can be almost equal to wet deposition in some areas (Lovett and Lindberg 1993), we assumed a uniform dry: wet deposition ratio of 3: 7 and then estimated the dry deposition rate from that of wet deposition.

For organic fertilizer N input, we considered human and animal manure, as well as crop residues returned to the soil. Previous studies (Liu 1991, Liu *et al* 2006) have shown that about 50% of animal and rural-human excreta are used for manure. Rates of excretion per head of animal per year have been estimated as 48.8 kg N for cattle (Wu 2005), 8.1 kg N for pigs, 5.7 kg N for sheep, 27.2 kg N for horses, and 15.5 kg N for mules and donkeys (Liu *et al* 2006); while for humans it is 5 kg N yr^{1} (Xing and Yan 1999). Data on rural human and animal population in each province was obtained from China Agriculture Yearbooks (see the supplementary material for details) and the Comprehensive Statistical Data and Materials on 55 Years of New China (Department of Comprehensive Statistics of the National Economy, 2005). The amount of N in crop residues was calculated using the residue/harvest ratio and the N concentration in crop residues from the published literature (see table S1 for data sources). The proportions of residues of major crops returned to the soil were also obtained from the literature (Gao *et al* 2002).

Changes in Soil N Storage

Changes in soil N content are hard to detect over a short period of time because the change is usually small compared to the large soil N pool. China conducted a national soil survey during 1979–82, and the main results were published in the China Soil Series I–VI (National Soil Survey Office, 1993, 1994, 1995, 1996). These six books contain descriptions (including soil series, location and land use) and some analytical data for 2473 typical soil profiles, among which 1509 soil profiles were from croplands throughout the country. The dataset has been used to estimate the soil organic carbon (SOC) storage of Chinese croplands (Wuet *al* 2003, Xie *et al* 2007). Data of total N content for soil depths of 0–100 cm are available for 1213 cropland soil profiles.

We conducted another sampling campaign in 2007–08, with soil profile samples (0–100 cm) of 1393 cropland sites across the country. Organic carbon and total N content along with other properties of the samples were analyzed. We have shown that, with these numbers of soil profiles, it is possible to detect statistically-robust changes in the average SOC content (0–100 cm) of Chinese croplands; the detailed methodology of the estimation process is described in Yan *et al* (2011). Using a similar method in the present study, we detected the changes in the average soil N content in Chinese croplands (0–100 cm) over the period 1980–2007. The following describes the statistical analysis method used to compare the changes in soil N content:

Soil N content does not usually follow a normal distribution. According to (Bernstein and Bernstein 1999), the confidence interval (CI) for the mean of such a population with large sample size (≥30) can be estimated as

$$\overline{x} \pm z_{\alpha/2}\frac{s}{\sqrt{n}}, \tag{2}$$

where $\overline{x}$, s and n are the mean, standard deviation and size of the sample, respectively; and $z_{\alpha/2}$ is the Z statistic of a critical level ( ).

The CI of the mean difference of two independent large samples (size ≥30) can be estimated as

$$\left(\overline{x}_1 - \overline{x}_2\right) \pm z_{\alpha/2}\sqrt{\frac{s_1^2}{n_1} + \frac{s_2^2}{n_2}}, \tag{3}$$

where $\overline{x}_1$ and $\overline{x}_1$ are the means of the two samples; s_1 and s_2 are the standard deviations of the two samples; and n_1 and n_2 are the sizes of the two samples.

The statistical significance of the mean difference of two large samples can be tested using the statistic

$$\frac{(\bar{x}_1 - \bar{x}_2)}{\sqrt{(s_1^2 / n_1) + (s_2^2 / n_2)}}, \qquad (4)$$

and comparing with a standard normal distribution.

The sample taken between 1979–82 and that taken in 2007–08 can be viewed as two independent samples, so we were able to detect if the average soil N content has changed during this time period.

Statistical Evaluation of the Relationship between Crop N Uptake and N Input

To test whether the increasing trend of crop N uptake is statistically significant, we fitted the data (crop N uptake, synthetic N input and total N input) with a simple linear regression model. On the basis of the regression statistics ($r^2 > 0.53$; $P < 0.001$, $n = 30$), we concluded that the increasing trend of crop N uptake with N input is robust and statistically valid. Here, $n = 30$ indicates data points of a given year for all provinces of mainland China except Shanghai.

RESULTS

Soil N Change, Crop N Uptake and Cropland N Input

The average soil N content of Chinese croplands (0–100 cm) was 0.738 g N kg^{-1} during the period 1979–82 and 0.777 g N kg^{-1}during 2007–08, representing a significant increase of 5.1% ($p < 0.02$) between the two periods. Therefore, there was no net removal of soil N by crops during the total period.

We calculated annual total crop N uptake and total N input to croplands in China for the period 1980–2010. The total crop N uptake doubled over the period, while the synthetic N fertilizer tripled and the total N input increased by 141%. Synthetic N fertilizer dominated total N input (table 1).

Table 1: Total crop N uptake and N input to croplands of China in selected years. Data for synthetic N application rate and crop yield were obtained from national statistics. Crop N uptake was calculated from crop yield, grain/straw ratio, and crop N content with parameters shown in table S1

Year	Crop N uptake (Tg N yr^{-1})	N input to croplands(Tg N yr^{-1})				Crop N uptake as a proportion of total N input
		Synthetic fertilizer	Organic fertilizer	Biological fixation	Atmospheric deposition	
1980	8.1 (62[a])	9.4 (72)	3.7 (28)	2.8 (21)	1.4 (11)	0.47
1985	10.5 (80)	12.6 (96)	4.4 (34)	2.6 (20)	1.9 (14)	0.49
1990	12.0 (92)	17.4 (133)	5.0 (38)	2.6 (20)	2.4 (18)	0.44
1995	13.4 (103)	22.3 (171)	6.0 (46)	3.0 (23)	2.6 (20)	0.40
2000	14.1 (108)	24.6 (189)	6.1 (47)	3.1 (24)	2.8 (22)	0.38
2005	15.2 (117)	26.6 (205)	6.6 (51)	3.0 (23)	3.1 (24)	0.39
2010	16.9 (139)	29.5 (242)	5.7 (47)	2.9 (24)	3.9 (32)	0.40

[a]The unit of data in brackets is kg N per hectare of arable land area.

Relationship between Crop N Uptake and N Input at the Provincial Scale

We further calculated total crop N uptake and synthetic N input to croplands on a yearly basis from 1980–2010 for each province of China. At this scale, there was a generally linear relationship between crop N uptake and synthetic N input rate (figure 1). For a given year, the crop N uptake is derived from synthetic fertilizer N applied in that year and from a variety of other sources (soil organic N, fertilizer N applied in previous years, N deposition from the atmosphere, biological N fixation, applied crop residue and manure). The slope of the line can be viewed as the average fraction of applied synthetic N taken up by crops in the year, while the intercept of the line can be viewed as the average N uptake by crops from all other sources. Clearly, with the increase in N input, the fraction of applied synthetic N taken up by crop within a year decreased, from around 0.5 in the early-1980s to around 0.3 in the late-2000s (figure 2). By definition, the slope of the regression line should be similar to the RE_N calculated by the N difference method in one-year experimental plots, and indeed the slopes for recent years were (<0.30, figure 2), agreeing well with the widely reported RE_N of major crops in China for 2001–05 (Zhang *et al* 2007). Figure 3 shows that the determinant coefficient between fertilizer application rate and crop N uptake (R^2 in figure 2) generally decreased with fertilizer application rate, indicating the loosening of the link between fertilizer input and crop N uptake. The rapid decreasing in determinant coefficient when national average fertilizer N application rate was above 190 kg N ha^{-1} probably indicates over fertilization at such rate (occurred in the 1990s). Similar phenomenon was revealed by Tian *et al* (2012) and as attributed to contributions of other factors such as genetic improvement to crop yield in addition to overuse of N fertilizer.

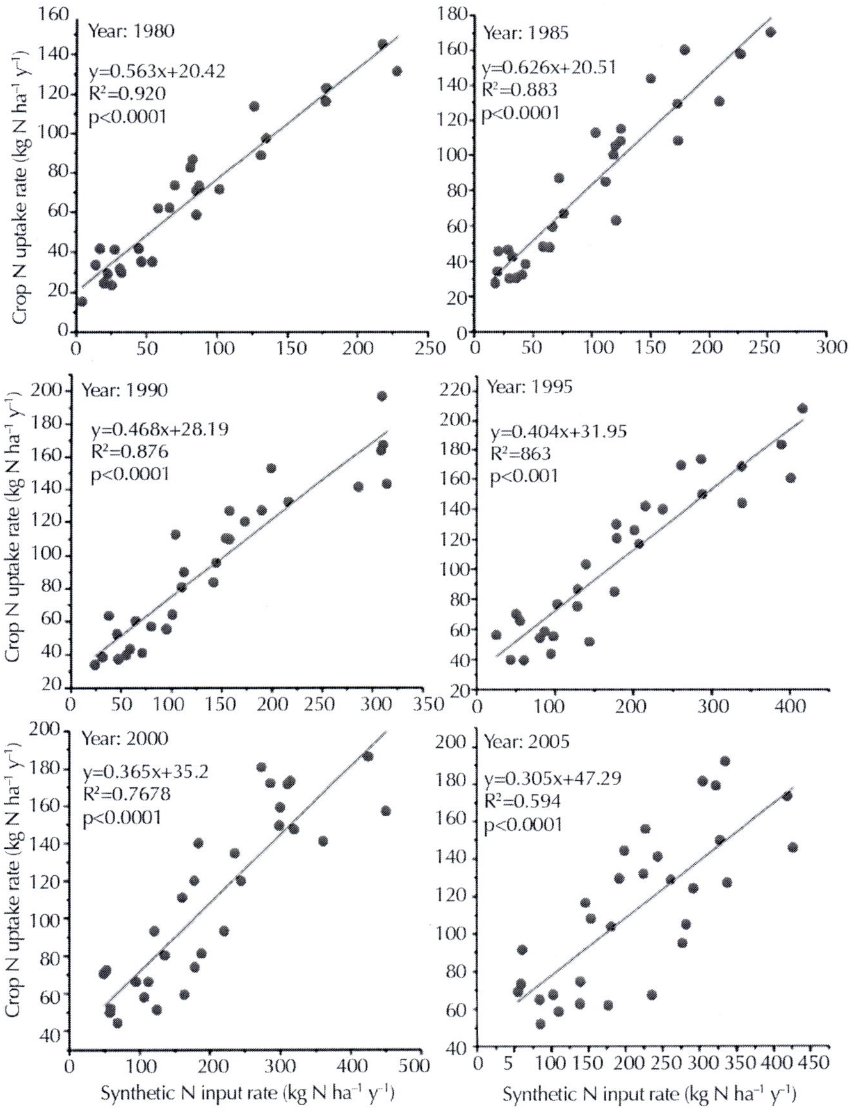

Figure 1: Relationship between synthetic fertilizer N application rate and crop N uptake for selected years. Data points of a given year are for all provinces of China except Shanghai. Shanghai had a much higher N application rate than other provinces, but its total use of synthetic N was negligible in comparison to the whole of China.

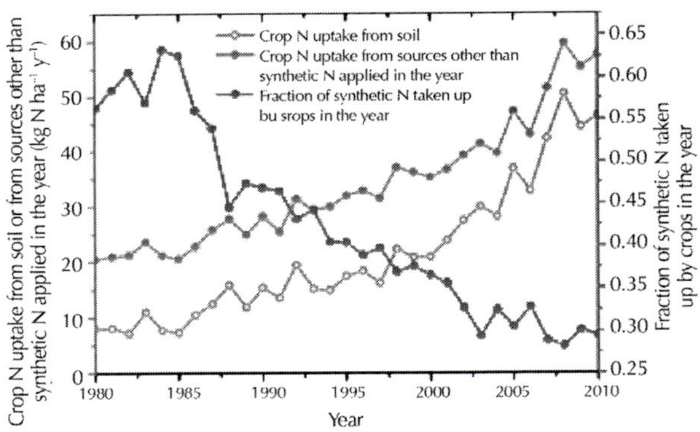

Figure 2: Changes in the average fraction of applied synthetic N taken up by crops in each year, as well as crop N uptake from soil or sources other than synthetic N applied that year.

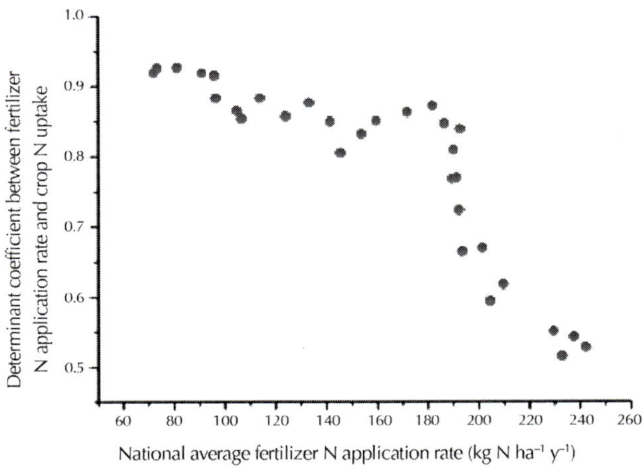

Figure 3: The change in determinant coefficient between fertilizer N application rate and crop N uptake with fertilizer application rate.

Crop N uptake from sources other than synthetic fertilizer applied to a site in a given year includes soil N, biologically-fixed N, atmospheric N deposition, and N in applied crop residues and manure. To distinguish soil N from other N sources, we regressed crop N uptake against total N input to croplands of different provinces in China for any given year (see figure 4 for selected years). Again, the slope of the regression line can be viewed as the fraction of total N input taken up by crops in the year, while the intercept of the line can be viewed as the average N uptake by crop from soils. The crop N uptake from soils increased continuously, from 8.0 kg N ha^{-1} yr^{-1} in 1980 to 46.4 kg N ha^{-1} yr^{-1} in 2010 (figures 2 and 4). This increase was obviously due to the residual effect of N input in previous years, especially synthetic fertilizer N since it dominated total N input.

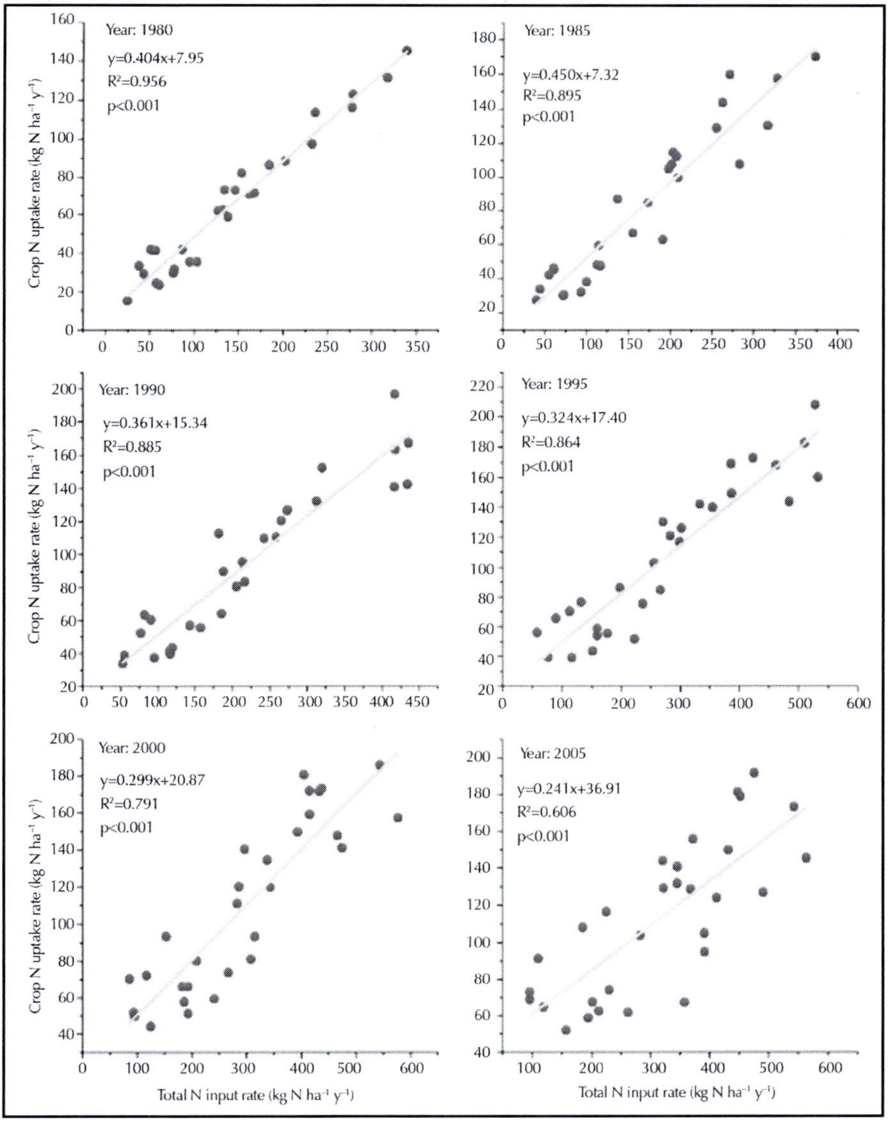

Figure 4: Relationship between total N input rate and crop N uptake for selected years. Data points of a given year are for all provinces of China except

Shanghai. Shanghai had a much higher total N input than other provinces, due to high synthetic N application rates, but its total use of synthetic N was negligible in comparison to the whole of China. See the supplementary material for the calculation of total N input.

DISCUSSION

Several studies have shown substantial increases in the average organic carbon content of Chinese cropland soils since the 1980s (Xie *et al* 2007, Piao *et al* 2009, Yu *et al* 2009). By comparing the data resulting from the national soil survey conducted in 1979–82 and soil samples collected in 2007–08 across the country, we found that the average C/N ratio of cropland soils did not change significantly, so we are confident that soil N has not depleted. For large-scale croplands, in the long-term, if soil N has not depleted, the N taken up by crops must all have come from N inputs, including synthetic fertilizer N, N in applied manure and crop residue, biologically-fixed N, and atmospheric N deposition. Then, the ratio between total crop N uptake and total N input is the average RE of the N input. The ratio between crop N uptake and total N input ranged between 0.35–0.50, averaging 0.42 (table 1for selected years), higher than the RE_N of fertilizer N obtained in field experiments (Zhu and Chen 2002, Zhang *et al* 2007).

Given that synthetic N is more easily available to crops than other N sources (Zhu 1998), the ratio of crop N uptake to total N input (blue line in figure 5) can be viewed as the minimum proportion of synthetic N that is eventually used by crops. The ratio of crop N uptake to synthetic fertilizer N input (red line in figure 5) can be viewed as the maximum proportion of synthetic N that is eventually used by crops in China through uptake in the year, environmental effects in the year, and accumulated residual effects. The real fraction of synthetic N used by crops is between the maximum and minimum, including the fraction of synthetic N that is used in the year and the residual effect. Since there was no net decrease in soil N, the N taken up by crops from soil can be viewed as the residual effect of N input in previous years. Here, we therefore propose calculating the average accumulative RE of synthetic N of a country/region as

$$\left(RE_{N-in-year} \times FN_{rate} + N_{soil} * FN_{rate} / TN_{rate}\right) / FN_{rate,} \quad (5)$$

which is simplified as

$$RE_{N-in-year} + N_{soil} / TN_{rate} \quad (6)$$

where $RE_{N\text{-in-year}}$ is the fraction of synthetic N taken up by crops in the year (the slope of the linear line in figure 1), FN_{rate} is the average rate of synthetic N input (kg N ha^{-1}), TN_{rate} is the average rate of total N input (kg N ha^{-1}), and N_{soil} is the N uptake by crops from soil (the intercept of the linear line in figure 4; in kg N ha^{-1}). Following this, the accumulative recovery efficiencies of synthetic N in China were estimated to have ranged between 0.40 and 0.68 during 1980–2010, averaging 0.51 and being around 0.43 in recent years (green line in figure 5). These estimated accumulative recovery efficiencies were 10%–46% higher than their corresponding recovery efficiencies in annual terms (blue line in figure 2), due to the residual effect of N.

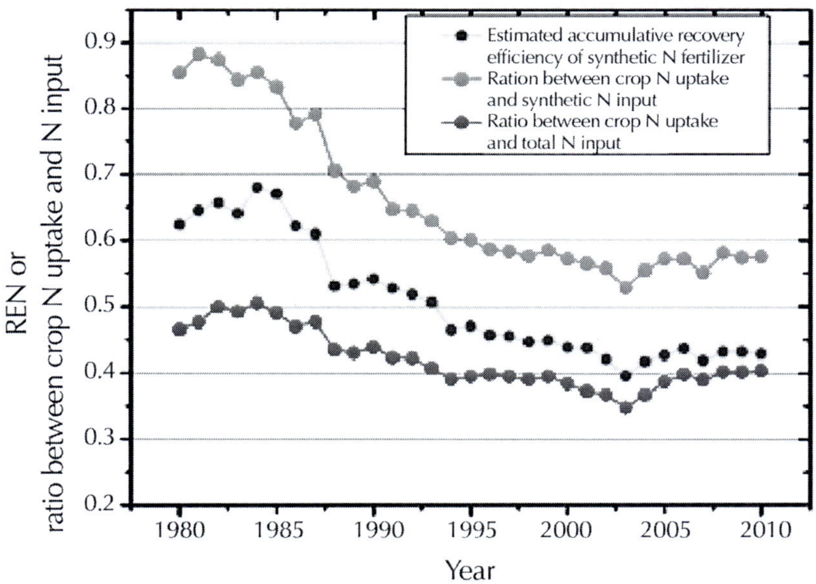

Figure 5: Estimated RE_{N}, as well as the ratio between crop N uptake and synthetic N input or total N input. Data represent the national average on an

annual basis.

At the end of the growing season, residual N from fertilizer can remain in the soil in the following forms: as mineral N, in roots, immobilized into the microbial biomass, or incorporated into other soil organic matter pools; and then become available for plant uptake during subsequent growing seasons (Dourado-Neto *et al* 2010). Although some ^{15}N-isotope experiments have shown little residual effect of fertilizer N (Timmons and Cruse 1991, Xu *et al* 1993), this effect may depend on the amount of N input. Continuous high N input to croplands of China has led to large amounts of soil N accumulation. For example, by undertaking 269 on-farm demonstration trials in the North China Plain, Cui *et al* (2013) showed that NO_3-N in the top 90 cm of the soil profile averaged 191 kg N ha^{-1} before the sowing of wheat, and 184 kg N ha^{-1} before the sowing of maize. Therefore, testing of soil NO_3–N became an essential step in fertilizer recommendation in the North China Plain (Cui *et al* 2008, 2013). In the semiarid areas of northwestern China, Zhao *et al* (2013) conducted a ^{15}N microplot experiment within long-term fertilization experiment plots and found that 7–15% of the ^{15}N-labeled fertilizer was recovered in the two subsequent crop seasons.

The significant residual effect of N fertilizer in Chinese croplands is also reflected by the large differences in RE_N obtained from long-term and short-term experiments. Results of the China National Soil Fertility and Fertilizer Efficiency Long-term Monitoring Network showed that RE_N values of the maize–wheat system (uplands) had a range of 0.56–0.72,calculated from the difference between NPK and PK treatments (Liu *et al* 2010), while the RE_N of the wheat–rice system (paddy fields) was around 0.40 (Shen *et al* 2007, Zhao *et al* 2010). In a series (hundreds) of short-term field experiments during 2001–05, Zhang *et al* (2007) showed RE_Nvalues of 0.283 for rice, 0.282 for wheat and 0.261 for maize. This increasing residual effect is also seen from the change in the yields of zero-N plots. For example, based on 7410 on-farm trials, Fan *et al* (2013) found that the yield of control plots (irrigated cereal-based cropping systems) was 0.73–1.76 Mg ha^{-1} higher in the 2000s than in the 1980s, and this was primarily due to the enhancement of soil-related factors since the yield of control plots of long-term experiments did not change significantly.

In addition to the significant residual effect, the increased soil N content translates to 6.8% of the total N input to croplands during 1980 and 2007. This suggests that the use efficiency of synthetic N is much higher and the loss rate is less than generally perceived.

Nevertheless, there remain many uncertainties in our estimation. For example, biologically-fixed N is an important source of total N input, but it has a wide range of rates reported in the literature. Furthermore, N deposition, especially dry deposition, can contribute to uncertainties in N input due to the scarcity of measurements. Nevertheless, synthetic N fertilizer is by far the dominant source of N input of croplands and our analysis clearly shows the large residual effect of fertilizer N in the crop production systems of China.

Though the accumulative recovery efficiencies of synthetic N in China were estimated to have ranged between 0.40 and 0.68 during 1980–2010, decades of fertilizer N overuse have contributed to soil, water, and air pollution. For example, ammonia and nitrous oxide emissions dominated by agricultural soils in China are the major contributions of heavy wet deposition and global warming (Reis *et al* 2009, Liu *et al* 2013), and fertilizer runoff accounted for 22–57% of the total nonpoint N pollution in large river basins in China (Ti and Yan 2013). Experimental evidence indicated that cutting 30–60% N would reduce more than 50% N loss without significantly diminishing crop yields in some intensive cropping systems in China (Ju *et al* 2009). Therefore the residual effect of previously applied N, as identified in this study, should be given full consideration in determining N application rate.

CONCLUSIONS

In China, due to the continuously high N input to crop production systems, the fertilizer N RE in annual terms is indeed low (below 30% in recent years), but the residual effect of fertilizer N has been increasing continuously, to the point that 40–68% of applied fertilizer is taken up sooner or later. Therefore, while there is scope to reduce N fertilizer use in China, the residual effect of applied fertilizer must be accounted for in any such effort.

Acknowledgments

This research was funded by the Chinese Academy of Sciences (Grant No. KZCX2-YW-GJ01).

REFERENCES

1. Asman W A H 1998 Factors influencing local dry deposition of gases with special reference to ammonia Atmos. Environ. 32 415–21
2. Bernstein S and Bernstein R 1999 Elements of Statistics II: Inferential Statistics (New York: McGraw-Hill)
3. Burns R C and Hardy R W F 1975 Nitrogen Fixation in Bacteria and Higher Plants (New York: Springer)
4. Cassman K G, Dobermann A and Walters D T 2002 Agroecosystems, nitrogen-use efficiency, and nitrogen management Ambio. 31 132–40
5. Cui Z L et al 2008 On-farm evaluation of the improved soil N(min)-based nitrogen management for summer maize in North China Plain Agron. J. 100 517–25
6. Cui Z L, Chen X P and Zhang F S 2013 Development of regional nitrogen rate guidelines for intensive cropping systems in China Agron. J. 105 1411–6
7. Department of Comprehensive Statistics of the National Economy 2005 Comprehensive Statistical Data and Materials on 55 Years of New China (Beijing: Statistics Press))
8. Dourado-Neto D et al 2010 Multiseason recoveries of organic and inorganic nitrogen-15 in tropical cropping systems Soil. Sci. Soc. Am. J. 74 139–52
9. Eurostat 2013 Nitrogen balance in agriculture Available online at: http://epp.eurostat.ec.europa.eu/statistics_explained/index.php/Nitrogen_balance_in_agriculture (Access 13/04/2014)
10. Fan M S, Lal R, Cao J, Qiao L, Su Y S, Jiang R F and Zhang F S 2013 Plant-based assessment of inherent soil productivity and contributions to China's cereal crop yield increase since 1980 PLoS One 8 e74617

11. Fowler D et al 2013 The global nitrogen cycle in the twenty-first century Phil. Trans. R. Soc. B 368 20130164
12. Galloway J N 1998 The global nitrogen cycle: changes and consequences Environ. Pollut. 102 15–24
13. Galloway J N et al 2004 Nitrogen cycles: past, present, and future Biogeochemistry 70 153–226
14. Galloway J N, Townsend A R, Erisman J W, Bekunda M, Cai Z C, Freney J R, Martinelli L A, Seitzinger S P and Sutton M A 2008 Transformation of the nitrogen cycle: recent trends, questions, and potential solutions Science 320 889–92
15. Gao X Z, Ma W Q and Ma C B 2002 Analysis on the current status of utilization of crop straw in China J Huazhong Agric Univ. 21 242–7 (in Chinese, with English abstract)
16. Goulding K W T, Bailey N J, Bradbury N J, Hargreaves P, Howe M, Murphy D V, Poulton P R and Willison T W 1998 Nitrogen deposition and its contribution to nitrogen cycling and associated soil processes New Phytol. 139 49–58
17. Green P A, Vorosmarty C J, Meybeck M, Galloway J N, Peterson B J and Boyer E W 2004 Pre-industrial and contemporary fluxes of nitrogen through rivers: a global assessment based on typology Biogeochemistry 68 71–105
18. Jensen L S and Schjoerring J K 2011 The European Nitrogen Assessment ed M A Sutton et al (Cambridge: Cambridge University Press) 32–61
19. Ju X T et al 2009 Reducing environmental risk by improving N management in intensive Chinese agricultural systems Proc. Natl. Acad. Sci. USA 106 3041–6
20. Krupnik T J, Six J, Ladha J K, Paine M J and Van Kessel C 2004 Agriculture and the Nitrogen Cycle: Assessing the Impacts of Fertilizer Use on Food Production and the Environment ed A R Mosier, J K Syers and J R Freney (Paris, France: Island Press) 193–207
21. Ladha J K, Pathak H, Krupnik T J, Six J and van Kessel C 2005 Efficiency of fertilizer nitrogen in cereal production: retrospects and prospects Adv. Agron. 87 85–156

22. Leip A, Britz W, Weiss F and Vries W 2011 Farm, land, and soil nitrogen budgets for agriculture in Europe calculated with CAPRI Environ. Pollut. 159 3243–53
23. Liu G L 1991 China Organic Fertilizer (Beijing: Agriculture Press) (in Chinese)
24. Liu J, Liu H, Huang S M, Yang X Y, Wang B R, Li X Y and Ma Y B 2010 Nitrogen efficiency in long-term wheat maize cropping systems under diverse field sites in China Field Crops Res. 118 145–51
25. Liu X J et al 2013 Enhanced nitrogen deposition over China Nature 494 459–62
26. Liu X L, Xu J X and Wang F H 2006 Estimation parameters of nitrogen balance in stock farming system of China Chinese J. Appl. Econ. 17 417–23 (in Chinese, with English abstract)
27. Liu X Y, He P, Jin J Y, Zhou W, Sulewski G and Phillips S 2011 Yield gaps, indigenous nutrient supply, and nutrient use efficiency of wheat in China Agron. J. 103 1452–63
28. Lovett G M and Lindberg S E 1993 Atmospheric deposition and canopy interactions of nitrogen in forests Can. J. For. Res. 23 1603–16
29. Lü C Q and Tian H Q 2007 Spatial and temporal patterns of nitrogen deposition in China: synthesis of observational data J. Geophys. Res. 112 D22S05
30. Ministry of Land and Resources of China 2008 Communique on Land and Resources of China 2007 (Available at http://www.gov.cn/gzdt/2008-04/17/content_947023.htm
31. Ministry of Land and Resources of China 2009 Communique on Land and Resources of China 2008 (Available at http://www.mlr.gov.cn/wszb/20090331bzzbhxdzzk/beijingziliao/200903/t20090331_687345.htm
32. Ministry of Land and Resources of China, National Bureau of Statistics of China and National Agricultural Census Office of China 1999 Communique of Main Data on Land Use Survey (Available at http://www.mlr.gov.cn/zwgk/tjxx/201108/t20110804_916414.htm
33. National Soil Survey Office 1993 China Soil Series vol I (Beijing: China Agricultural Press) (in Chinese)

34. National Soil Survey Office 1994 China Soil Series vol II–III (Beijing: China Agricultural Press) (in Chinese)
35. National Soil Survey Office 1995 China Soil Series vol IV–V (Beijing: China Agricultural Press) (in Chinese)
36. National Soil Survey Office 1996 China Soil Series vol VI (Beijing: China Agricultural Press) (in Chinese)
37. Ohara T, Akimoto H, Kurokawa J, Horii N, Yamaji K, Yan X and Hayasaka T 2007 An Asian emission inventory of anthropogenic emission sources for the period 1980–2020 Atmos. Chem. Phys. 7 4419–44
38. Piao S L, Fang J Y, Ciais P, Peylin P, Huang Y, Sitch S and Wang T 2009 The carbon balance of terrestrial ecosystems in China Nature 458 1009–U82
39. Qiao J, Yang L Z, Yan T M, Xue F and Zhao D 2012 Nitrogen fertilizer reduction in rice production for two consecutive years in the Taihu lake area Agric. Ecosyst. Environ. 146 103–12
40. Raun W R and Johnson G V 1999 Improving nitrogen use efficiency for cereal production Agron. J. 91 357–63
41. Reis S, Pinder R W, Zhang M, Lijie G and Sutton M A 2009 Reactive nitrogen in atmospheric emission inventories Atmos. Chem. Phys. 9 7657–77
42. Roger P A and Ladha J K 1992 Biological N2 fixation in wetland rice fields—estimation and contribution to nitrogen balance Plant Soil 141 41–55
43. Shen M X, Yang L Z, Yao Y M, Wu D D, Wang J G, Guo R L and Yin S X 2007 Long-term effects of fertilizer managements on crop yields and organic carbon storage of a typical rice-wheat agroecosystem of China Biol. Fertil. Soils 44 187–200
44. Smil V 1999 Nitrogen in crop production: an account of global flows Global Biogeochem. Cycles 13 647–62
45. Ti C P, Pan J J, Xia Y Q and Yan X Y 2012 A nitrogen budget of mainland China with spatial and temporal variation Biogeochemistry 108 381–94
46. Ti C P and Yan X Y 2013 Spatial and temporal variations of river nitrogen exports from major basins in China Environ. Sci. Pollut. Res. 20 6509–20

47. Tian H et al 2012 Food benefit and climate warming potential of nitrogen fertilizer uses in China Environ. Res. Lett. 7 044020
48. IOPscience
49. Tilman D 1999 Global environmental impacts of agricultural expansion: the need for sustainable and efficient practices Proc. Natl. Acad. Sci. USA. 96 5995–6000
50. Timmons D R and Cruse R M 1991 Residual N-15 recovery by corn as influenced by tillage and fertilization method Agron. J. 83 357–63
51. Vitousek P M, Aber J D and Howarth R W 1997 Human alteration of the global nitrogen cycle: sources and consequences Ecol. Appl. 73 737–50
52. Wu H B, Guo Z T and Peng C H 2003 Distribution and storage of soil organic carbon in China Global Biogeochem. Cycles 17 1048
53. Wu S X 2005 The spatial and temporal change of nitrogen and phosphorus produced by livestock and poultry& their effects on agricultural no-point pollution in China PhD Thesis The Chinese Academy of Agricultural Sciences (in Chinese, with English abstract)
54. Xie Y X 2006 Nitrogen originating from environment in paddy ecosystem under anthropogenic influences PhD Thesis Institute of Soil Science, Chinese Academy of Sciences (in Chinese)
55. Xie Z B, Zhu J G, Liu G, Cadisch G, Hasegawa T, Chen C M, Sun H F, Tang H Y and Zeng Q 2007 Soil organic carbon stocks in China and changes from 1980s to 2000s Global Change Biol. 13 1989–2007
56. Xing G X and Yan X Y 1999 Direct nitrous oxide emissions from agricultural fields in China estimated by the revised 1996 IPPC guidelines for national greenhouse gases Environ. Sci. Policy 2 355–61
57. Xu Z H, Myers R J K, Saffigna P G and Chapman A L 1993 Nitrogen fertilizer in leucaena alley cropping. II. residual value of nitrogen fertilizer and leucaena residues Fert. Res. 34 1–8
58. Yan W J, Zhang S, Sun P and Seitzinger S P 2003 How do nitrogen inputs to the Changjiang basin impact the Changjiang river

nitrate: a temporal analysis for 1968-1997 Global Biogeochem. Cycles 17 1091

59. Yan X Y, Cai Z C, Wang S W and Smith P 2011 Direct measurement of soil organic carbon content change in the croplands of China Global Change Biol. 17 1487–96
60. Yang R, Hayashi K, Zhu B, Li F Y and Yan X Y 2010 Atmospheric NH(3) and NO(2) concentration and nitrogen deposition in an agricultural catchment of eastern China Sci. Total Environ. 408 4624–32
61. Yearbook E B o C A 1981–2011 1981–2011 China Agriculture Yearbook (Beijing: China Agricultural Press) (in Chinese)
62. Yu Y Y, Guo Z T, Wu H B, Kahmann J A and Oldfield F 2009 Spatial changes in soil organic carbon density and storage of cultivated soils in China from 1980 to 2000 Global Biogeochem. Cycles 23 GB2021
63. Zhang F S, Cui Z L, Wang J Q, Li C J and Chen X P 2007 Current status of soil and plant nutrient management in China and improvement strategies Chin. Bull. Bot. 24 687–94 (in Chinese with English abstract)
64. Zhao B Q et al 2010 Long-term fertilizer experiment network in China: crop yields and soil nutrient trends Agron. J. 102 216–30
65. Zhao W, Lliang B, Yang X and Zhou J 2013 Effects of long term different fertilization on the fate of residual fertilizer N in a what -maze rotation system Sci. Agric. Sin. 46 1628–34 (in Chinese with English abstract)
66. Zhu Z L 1998 Soil Fertility of China ed X M Shen (Beijing: China Agriculture Press) 484
67. Zhu Z L and Chen D L 2002 Nitrogen fertilizer use in China—Contributions to food production, impacts on the environment and best management strategies Nutr. Cycling Agroecosyst. 63 117–27
68. Zhu Z L and Wen Q X 1992 Nitrogen in Soil of China (Nanjing, China : Jiangsu Science and Technology Publishing House) (in Chinese)

chapter 12

Overcoming Weed Management Challenges in Cover Crop–Based Organic Rotational No-Till Soybean Production in the Eastern United States

Steven B. Mirsky, Matthew R. Ryan, John R. Teasdale, William S. Curran, Chris S. Reberg-Horton, John T. Spargo, M. Scott Wells, Clair L. Keene, and Jeff W. Moyer*

*First and third authors: Research Ecologist and Research Plant Physiologist, Sustainable Agricultural Systems Laboratory, USDA-ARS, Beltsville, MD 20705; Second author: Assistant Professor, Department of Crop and Soil Sciences, Cornell University, Ithaca, NY 14850; fourth and eighth authors: Professor and Graduate Research Assistant, Department of Plant Science, The Pennsylvania State

University, University Park, PA 16802; fifth and seventh authors: Associate Professor and Graduate Research Assistant, Department of Crop Science, North Carolina State University, Raleigh, NC 27695; sixth author: Assistant Professor, Stockbridge School of Agriculture, University of Massachusetts, 682 North Pleasant St., Amherst, MA 01003; ninth author: The Rodale Institute, Kutztown, PA 19530.

ABSTRACT

Cover crop–based organic rotational no-till soybean production has attracted attention from farmers, researchers, and other agricultural professionals because of the ability of this new system to enhance soil conservation, reduce labor requirements, and decrease diesel fuel use compared to traditional organic production. This system is based on the use of cereal rye cover crops that are mechanically terminated with a roller-crimper to create in situ mulch that suppresses weeds and promotes soybean growth. In this paper, we report experiments that were conducted over the past decade in the eastern region of the United States on cover crop–based organic rotational no-till soybean production, and we outline current management strategies and future research needs. Our research has focused on maximizing cereal rye spring ground cover and biomass because of the crucial role this cover crop plays in weed suppression. Soil fertility and cereal rye sowing and termination timing affect biomass production, and these factors can be manipulated to achieve levels greater than 8,000 kg ha^{-1}, a threshold identified for consistent suppression of annual weeds. Manipulating cereal rye seeding rate and seeding method also influences ground cover and weed suppression. In general, weed suppression is species-specific, with early emerging summer annual weeds (e.g., common ragweed), high weed seed bank densities (e.g. $>$ 10,000 seeds m^{-2}), and perennial weeds (e.g., yellow nutsedge) posing the greatest challenges. Due to the challenges with maximizing cereal rye weed suppression potential, we have also found high-residue cultivation to significantly improve weed control. In addition to cover crop and weed management, we have made progress with planting equipment and planting density for establishing soybean into a thick cover crop residue. Our current and future research will focus on integrated multitactic weed management, cultivar selection, insect pest suppression, and nitrogen management as part of a systems approach to advancing this new production system.

INTRODUCTION

Soil and water conservation is essential to the long-term sustainability of crop production systems. Over the past 20 yr, no-tillage management has dramatically increased the ability to conserve soil resources (Franzluebbers 2002, 2005; Pesant et al. 1987; Spargo et al. 2008), while simultaneously reducing labor and production costs (Parsch et al. 2001). However, overreliance on herbicides has led to the selection of herbicide-resistant weeds, which has in some cases more than doubled weed management expenses (Price et al. 2011). As a result, soil conservation using no-till practices has constrained long-term weed management efficacy and sustainability in conventional agriculture. In contrast, traditional organic grain production requires extensive tillage that integrates diverse physical and cultural tactics, reducing the opportunity for weeds to develop resistance to a given tactic. Weed management in organic grain crop production typically involves full-inversion moldboard plow or disc tillage, cultivation with a tine weeder or rotary hoe, and interrow cultivation. On the other hand, organic cropping systems typically use cover crops, return crop residues, and use C-rich fertility amendments and crop diversification, which offset the negative impacts of tillage (Teasdale et al. 2007). An increasing number of growers are interested in developing reduced-tillage systems that integrate the soil-conserving and labor-saving features of conventional no-tillage systems with the soil building practices used in organic production. In this summary of our recent research, we describe advancements in the development of a cover crop–based, organic rotational no-till soybean production system that balances the trade-offs between soil conservation and weed management.

We first introduce our region and discuss the critical role of cover crops in this system and how they must be manipulated to maximize weed suppression. We then discuss the necessary equipment for optimizing soybean establishment and multitactic weed management with an emphasis on high-residue cultivation. We conclude with a broader discussion on the integration of this soybean production system into existing organic grain rotations.

Cover Crop–Based Organic Rotational No-Till System Production and Ecosystem Services

In response to a rising interest for reducing tillage in organic crop production, growers and researchers have developed an organic crop and soil management system that has since been termed *cover crop–based organic rotational no-till* (Mirsky et al. 2012). This approach relies on mature cover crop residue to be used as mulch in place of standard weed management tactics such as herbicides and cultivation. In the cover crop–based organic rotational no-till system, winter annual cover crops are controlled in the spring using a roller-crimper, which provides in situ surface mulch that physically suppresses weeds. The cash crop is then no-till planted into the cover crop mulch. However, tillage is typically used prior to seeding the cover crop to optimize establishment and to control perennial weeds; thus it is a rotational tillage system.

Cereal rye has been the primary cover crop used in conjunction with soybean (Berstein et al. 2011; Davis 2010; Delate et al. 2012; Moore et al. 1994). This cover crop has been found to be the most reliable winter annual cover crop for providing a wide range of ecological services in the eastern region (Shipley et al. 1992). Cereal rye is one of the earliest flowering species, extremely cold tolerant, produces the most biomass, and has rapid emergence (Clark 2007). Furthermore, cereal rye has been coupled with soybean for its residue persistence and flexible establishment date (Mirsky et al. 2011; Ruffo and Bollero 2003).

The perceived inability to use no-till for organic crop production has been suggested as a key factor limiting the adoption of organic practices by conventional producers (Cavigelli et al., 2013). Labor required for organic grain production is another major constraint limiting the amount of organically managed cropland. Effective strategies for reducing tillage in organic systems could serve to attract more farmers to adopt organic production and reduce current organic feed grain shortages. Integrating no-till practices can potentially improve the sustainability of organic grain production by reducing diesel fuel use and labor requirements. In a recent analysis of energy use in a 3-yr organic corn–soybean–wheat rotation, cover crop–based rotational no-till required 27% less diesel fuel and 31% less labor than traditional organic management (Mirsky et al. 2012; Ryan 2010). This reduction in diesel fuel use and labor is largely a result of eliminating seedbed preparations (tillage, disking, and cultipacking) and common cultivation approaches (rotary hoe or tine-weeders at crop establishment followed by interrow cultivators) used for weed control in traditional tillage-based organic corn and soybean production.

Winter annual cover crops provide numerous ecosystem services that improve crop performance, nutrient cycling, and overall cropping system function (Snapp et al. 2005). Cover crops protect surface- and groundwater quality by decreasing soil erosion, nitrogen (N) leaching, and phosphorus (P) runoff (Adeli et al. 2011; Qi and Helmers 2009). Using cover crops to improve N retention through scavenging and biological N fixation can lower fertilizer costs (Dabney et al. 2010. Increased N conservation, addition, or both with cover crops reduces the reliance of organic cropping systems on animal manure thereby reducing excessive P loading. Cover crops can also provide habitat for pollinators (Decourtye et al. 2010) and other beneficial insects, such as weed seed predators (Ward et al. 2011). Furthermore, the cover crop–based organic rotational no-till system relies on terminating cover crops at anthesis, which can result in one to two orders of magnitude greater root and shoot biomass relative to cover crops grown in traditional organic or conventional cropping systems. Despite these benefits, cover crop performance is of utmost importance because of the critical role cover crops play in weed suppression. It is also important to note that the services provided by cover crops are completely transferable to conventional systems, but their value may be reduced given the broader range of selective pest and precision fertility management options that are available to conventional growers.

Corn and soybean are the primary commodity crops that have been evaluated in the cover crop–based organic rotational no-till system in the central and eastern United States. While both crops can produce competitive yields in this system, the performance of corn has been more variable than soybean (Delate et al. 2012; Mirsky et al. 2012; Reberg-Horton et al. 2012). Greater success with soybean has been attributed to several factors. First, both crops are vulnerable to reduced plant populations due to poor seed placement and seed and seedling feeding herbivores; however, soybean has greater plasticity than corn and can more effectively compensate for stand losses. Second, soybean is a legume and does not have the same nitrogen management (i.e., source, timing, and placement) constraints as corn. Third, grass cover crops like cereal rye generally attain greater dry matter and produce a better weed suppressive mulch than legume cover crops used with corn.

REGIONAL PERSPECTIVE AND CONSTRAINTS

Agriculture in the eastern region of the United States includes an array of relatively small farms averaging 73 ha per farm (National Agriculture Statistics Service [NASS] 2011). This region produces a diverse range of commodities in integrated and decoupled crop and livestock operations. Approximately one third of the landscape is used for agriculture, with animal production representing a significant segment of the agricultural industry (U.S. Environmental Protection Agency 2007). For example, in the Northeast alone there are more than 17,200 dairy farms and 1.4 million milking cows (NASS 2011; Winsten et al. 2010). The region as a whole is a net importer of nutrients in the form of grain for livestock feed and fertilizer for crop production (Anser and Townsend 1997). Arable land in the region spans five physiographic provinces, which vary considerably in both topographic features and climate. As a result, soils in the region range from droughty to poorly drained, are shallow and highly weathered, and have low organic matter content and low fertility compared to midwestern soils. Droughty summer conditions are more problematic in the sandy coastal plain soils than in the heavier soils of the Appalachian Plateau and Ridge and Valley physiographic provinces. These droughty conditions are also problematic when moving on a north to south gradient. Furthermore, droughty summers reduce yield potential, crop growth, and subsequently nutrient cycling. Due to steep topography and high precipitation, agricultural production in the eastern United States is prone to erosion and nutrient loading in sensitive estuarine environments such as the Chesapeake Bay. As a result, soil conservation is a priority in the region. Research, farmer innovation, and monetary incentives from federal programs have resulted in considerable interest in using cover crops to, in part, address the production challenges mentioned above. Cover crops have potential to enhance soil conservation in the region without impacting crop performance since spring soil moisture is typically not limiting (annual precipitation is 760 to 1012 mm, evenly distributed throughout the year). Furthermore, cover crop residues can increase crop tolerance to droughty conditions by lowering surface moisture evaporation.

OPERATIONALIZING COVER CROP MANAGEMENT

Although using cover crop residue as mulch to suppress weeds in no-till planted crops has been researched for decades, this approach was not considered to be feasible by organic grain growers until the development of the cover crop roller-crimper. This technology was first developed in South America (Derpsch et al. 1991). Early investigations into roller-crimper technology in the United States was conducted by USDA-ARS researchers in Auburn, AL (Ashford and Reeves 2003; Kornecki et al. 2006), and popularized by the Rodale Institute in Kutztown, PA. The primary manufacturer of the roller-crimper has been I&J Manufacturing of Gap, PA. Blueprints for the roller-crimper, which is made from a steel cylinder with metal slats welded perpendicular to the cylinder and arranged in a chevron pattern to reduce vibration, were also made available online for free, furthering the availability of the tool (seehttp://www.rodaleinstitute.org/no-till_revolution). This relatively inexpensive tool enables organic growers to terminate cover crops mechanically, without the use of herbicides. Several factors contribute to interest in roller-crimpers compared with other methods of terminating cover crops, such as mowing. Roller-crimpers can be operated at faster speeds and do not require an energy-intensive power take-off (PTO) drive, and thus require less fuel and labor. Weed suppression provided by cover crops terminated with the roller-crimper can also be greater than that provided by cover crops terminated with mowers. This is because mowers unevenly distribute cover crops residue and the residue decomposes faster after maceration (Creamer and Dabney 2002). More recently, engineers at USDA-ARS in Auburn, AL, have developed new roller-crimpers for both grain and raised-bed vegetable production (Kornecki et al. 2009).

Mechanical control of cereal rye is growth-stage dependent with greater than 90% control typically observed when rolled at the milk stage or later (Kornecki et al. 2006; Mirsky et al. 2009). Ineffective cereal rye termination will result in recovery, regrowth, or both and potential crop competition as well as viable seed production, which can contaminate winter cereal grain cash crops in the rotation. The direction of planting the cash crop must correspond to the direction of rolling to minimize hair-pinning and wrapping of residues and row cleaners (Kornecki et al. 2009). In addition, the rye should not be rolled in the same direction it was seeded because this can reduce the uniformity of soil coverage and the soybean planter units run the risk of lining up in the rows of rye reducing effective seed placement.

WEED SUPPRESSION FROM CEREAL RYE

The necessity of using cover crops for weed suppression in reduced-tillage organic cropping systems highlights the importance of understanding the factors that influence this process. Cereal rye has been studied extensively over the last three decades for the physical and chemical attributes that contribute to its weed-suppressive capabilities (Barnes and Putnam 1983; Burgos et al. 1999;Mirsky et al. 2011; Teasdale and Mohler 1993). Under typical cover cropping practices across the eastern United States, cereal rye usually produces biomass levels of approximately 4,000 to 6,000 kg ha^{-1} when grown to maturity without an N fertilizer application, and as high as 10,000 to 12,000 kg ha^{-1} with optimal management (Mirsky et al. 2011; Reberg-Horton et al. 2012). Conventional growers often terminate rye at earlier growth stages when considerably less biomass has accumulated.

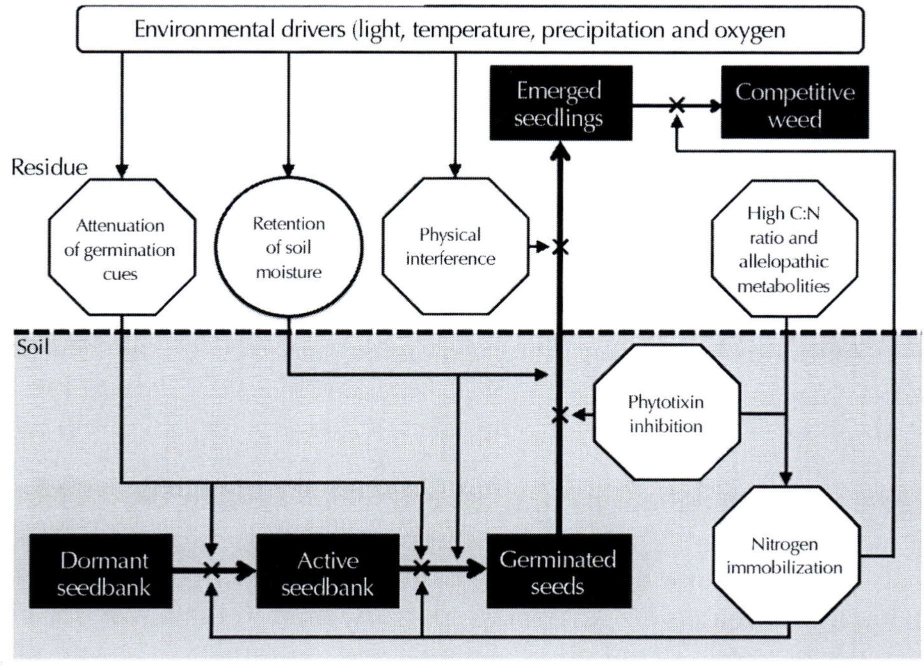

Figure 1: Schematic depicting the mechanisms for suppression or enhancement of weed germination and emergence by mulch from a cereal rye cover crop. Heavy solid arrows indicate weed life stage transitions. The lighter solid arrows indicate mulch induced factor effects that either inhibit or promote life stage transitions. Black rectangles connected by arrows represent weed life stages from dormant seeds to emerged seedlings. Octagons represent effects that inhibit weed life stage transitions at the point marked with an "X." Circles represent effects that enhance emergence at the points designated by the attached arrow.

Research with a range of surface rye mulch rates has shown that residue must be present in amounts substantially higher than these typical levels to effectively suppress annual weeds (Teasdale and Mohler 2000). For example, greater than 75% inhibition of the emergence of most annual weeds was obtained only when rye mulch biomass exceeded 8,000 kg ha^{-1} dry weight and mulch thickness exceeded 10 cm. The primary mechanisms by which cereal rye residues inhibit germination and emergence are by (1) attenuating environmental cues that break seed dormancy, (2) physically interfering with the emergence process, and (3) releasing phytotoxic compounds (Figure 1). Teasdale and Mohler (1993) showed that cereal rye residue can limit light and temperature cues that weed seeds often require for initiation of germination. Seeds that do germinate under high levels of mulch (>8,000 kg ha^{-1}) are inhibited primarily by physical interference of the mulch materials with the upward movement of emerging seedling and with the downward penetration of light to seedlings, whereby seed nutrient reserves are exhausted before seedlings reach sufficient light to become established. Release of phytotoxic compounds from cereal rye residue has been shown to inhibit weed germination and growth processes of many plant species, and the benzoxazinoid group of compounds have been identified as the most active compounds in cereal rye (Barnes and Putnam 1987; Macías et al. 2005; Reberg-Horton et al. 2005). Recent research suggests that benzoxazinoid compounds are not present in soil for more than 2 wk after rye termination and are found at concentrations too low to account for weed suppression (Rice et al. 2012). Thus, allelopathic effects that may occur are most likely caused by compounds other than benzoxazinoids. That said, physical rather than allelopathic influences probably predominate when mature cereal rye is terminated and used as a surface mulch (Teasdale et al. 2012).

The impact of cereal rye on soil inorganic N availability and the subsequent effects on weeds have not been as well characterized as the physical and allelochemical effects. The high biomass levels and residue quality (high C:N ratio, lignin, and hemicellulose content) of cereal rye results in very low inorganic N released to soil during residue decomposition and could contribute significantly to weed suppression by rye residue (Figure 1). Most nonleguminous weeds are highly responsive to N (Blackshaw et al. 2004; Henson and Jordan 1982; Tungate et al. 2006), and therefore manipulation of soil inorganic N levels could serve as a tactic for reducing weed competition in a cereal rye–mulched soybean cash crop. Frey et al. (2000) demonstrated (via ^{15}N tracer) fungal mediated upward movement of soil N from below the soil surface into the rye mulch. Research on exploiting niche differences and manipulating soil C:N ratio as a weed suppression tool is in its infancy, but some research suggests that it is effective and has multiple benefits (Phelan et al. 2008; Whitehouse et al. 2009). In a study of rolled rye vs. no-till plots without a cover crop mulch, soil inorganic N from plots with cereal rye was significantly lower in the surface to 10-cm throughout the duration of the growing season (Wells 2010). Immobilization of N played an important role in the continued suppression of available soil N, whereby available mineralized N from soil organic matter was quickly immobilized via fungi and bacteria. Beyond exhibiting classical N deficiency symptoms, weed shoot tissue N concentration was lower in the cereal rye–mulched treatments when compared to their no rye-mulch counterparts (Wells 2010). These results suggest that manipulating soil N with cereal rye mulch may play a larger role in weed suppression than previously thought.

Weed species vary widely in their sensitivity to cover crop residue, and the degree of sensitivity is mediated by soil edaphic features. Therefore, weed community structure and seed bank density will determine the success of cover crop–based weed management. Annual weed species with relatively smaller seed sizes are more sensitive to suppression by surface residue than larger seeded species (Mohler and Teasdale 1993; Teasdale and Mohler 2000).Mohler (1996) suggested that surface residue would inherently favor crops that have seed that is one to three orders of magnitude larger than seed of annual weeds. This selectivity between crops and weeds applies equally to differences in seed size among weed species and can be accounted for by greater nutrient and energy reserves in the endosperm of larger seeds to facilitate penetration of thicker mulch. In addition, larger seeds with greater energy and nutrient reserves may have greater metabolic capacity to detoxify allelopathic compounds or overcome a low N environment than smaller seeds. Perennial weeds with relatively large reserves and lower dormancy requirements than annual weeds can also be highly insensitive to or even stimulated by cover crop residue (Mirsky et al. 2011). Thus, regardless of the mass and thickness of residue achieved, weed species with little sensitivity to cover crop residue are likely to proliferate in a cover crop–based weed management system.

FACTORS DRIVING CEREAL RYE BIOMASS AND THEIR INTERACTIONS WITH WEEDS

Cover crop–based weed suppression, as described, is greatly affected by biomass levels. Teasdale and Mohler (2000) concluded that cereal rye does not typically reach biomass thresholds necessary for consistent weed suppression and that mulches for weed suppression have primarily been used in vegetable production systems in which straw mulch can be augmented. In response to this work, numerous studies have examined agronomic management practices for the purpose of enhancing cereal rye biomass levels, including planting and termination date timing (Ashford and Reeves 2003; Mirsky et al. 2011; Nord et al. 2011), seeding rate (Boyd et al. 2009; Ryan et al. 2011a) and soil fertility (Mirsky et al. 2012; Ryan et al. 2011a). The effects of these agronomic practices on cereal rye biomass and subsequent weed suppression (Table 1) are discussed below.

Table 1: Cultural practices that can be used to increase cereal rye biomass production

Practice	Benefit	Drawback
Select high-biomass cultivar	Does not require management modifications	Availability of seed may be limited
Sow earlier in fall	Increased nutrient scavenging	Need to wait until preceding crop is harvested
Terminate later in spring	Improved control of cereal rye with roller-crimper	Reduced soybean yield potential
Increase sowing rate	May provide greater ground cover and weed suppression	Increased seed costs and response limited to low initial rate
Supplement fertility	May provide opportunity to spread manure	Increased costs and gives weeds competitive advantage over soybean
Irrigate	May improve soil microbial activity and ability to plant soybean	Infrastructure needed and response limited to dry conditions when water may need to be conserved

Management Timing

Since cereal rye phenology is tightly regulated by thermal time (Mirsky et al. 2009), fall planting and spring termination of cereal rye will influence growth stage and biomass production. Therefore, management timing of cover crops can be used to maximize cereal rye biomass production. For example, in Pennsylvania, cereal rye planted at 10-d intervals from late August to mid-October and terminated at 10-d intervals from May 1 through May 30 ranged in biomass from 1,615 to 12,600 kg ha^{-1}. When cereal rye biomass accumulation was modeled on a thermal time basis (growing degree days), a 5–degree day difference in the fall was equivalent to a 1–degree day in the spring regarding rye biomass accumulation (Mirsky et al. 2011). After stem elongation (Zadoks growth stage 30), cereal rye can accumulate dry matter at a rate of 200 kg $ha^{-1}d^{-1}$ under optimal growing conditions. Within this same experiment, total weed density decreased with increasing cereal rye biomass, which was influenced by both cereal rye seeding and termination date. Early-emerging summer annual weeds such as common lambsquarters (*Chenopodium album* L.) were less effectively controlled because their emergence is synchronized with lower cereal rye biomass levels (Mirsky et al. 2011). Late-emerging summer annual weeds tend to have emergence periods synchronized with maximum cereal rye biomass and are therefore more effectively controlled. Finally, perennial weeds were not affected by cereal rye mulch termination timing or biomass level, a relationship well documented in the literature (Facelli and Picket 1991; Mirsky et al. 2011; Mohler and Teasdale 1993).

These species-specific responses of weeds to cover crop management timing prompted closer evaluation of this relationship. In 2008 and 2009, follow-up trials were tested across two cereal rye sowing dates and three cereal rye termination dates. Seeds of three summer annual weed species with known differences in emergence periodicity were added to microplots to evaluate the interplay between cereal rye termination date and weed emergence periodicity. Common ragweed, giant foxtail (*Setaria faberi* Herrm.), and smooth pigweed (*Amaranthus hybridus* L.) seeds were each added to microplots at 100, 450, or 1,050 seeds m^{-2} the previous fall after rye establishment. As hypothesized, delaying cover crop termination reduced common ragweed biomass when pooled over cereal rye sowing dates, which was a result of increased cereal rye biomass with later termination. This finding supported previous results (Mirsky et al. 2011) as well as the notion of delaying termination to maximize weed suppression. However, when we controlled for cereal rye biomass and compared early sowing and early termination to late sowing and late termination, total weed biomass was reduced with earlier cereal rye termination. In both years, these pairs of treatments resulted in approximately 8,000 to 9,000 kg ha^{-1} cereal rye biomass at termination. When pooled over years and weed seed densities, weed biomass was reduced by approximately 44% when cereal rye was sown and terminated early, compared to late. In both years and across all treatments, common ragweed was the dominant weed species. Cereal rye ground cover in the spring increases proportionately with earlier sowing in the fall, which also results in earlier maturation in the spring. The combined effects of increasing early spring ground cover and early termination of cereal rye appears to enhance weed suppression. Thus, because the roller-crimper does not control emerged weed seedlings, it appears that preempting emergence of common ragweed and other early-emerging species by planting cereal rye early and terminating relatively early can be beneficial if cereal rye biomass and control with the roller-crimper are not compromised.

Increasing Cereal Rye Seeding Rate

Results from a factorial experiment comparing different seeding rates that was conducted in Maryland and Pennsylvania in 2008 and 2009 showed that increasing cereal rye seeding rates could effectively increase weed suppression. When averaged across all site-years, increasing cereal rye seeding rate from 90 to 210 kg seed ha^{-1} significantly decreased weed biomass by 31% (Ryan et al. 2011a). Interestingly this increase in weed suppression was not due to increased cereal rye biomass at termination, but rather was attributed to increased ground cover early in the spring. These results prompted experimentation with several modifications to our cereal rye seeding approach for cover crop–based organic rotational no-till soybean. First, these results confirmed that using relatively high cereal rye seeding rates is justified despite the lack of effect on cereal rye biomass. Second, they led to the practice of using a dual seeding method approach in which a portion of the seed is broadcast and the remainder is drill seeded. The drill disturbs the soil surface and results in soil being moved into the interrow region. This strategy is thought to increase ground cover by filling interrow spaces with broadcasted seed, providing a balance between the consistency achieved with drill-seeding and the soil coverage achieved with broadcast seeding. Although the potential benefits of this approach has not been quantified, several research programs have adopted it as a best management practices for cereal rye establishment. The additional seed and tractor costs associated with dual seeding has to be weighed against potential improvements in weed suppression.

Increasing Soil Fertility

Cereal rye biomass production can be significantly constrained by soil fertility levels, particularly available N. At anthesis, shoot N concentration of cereal rye is typically between 9 and 11 g N kg^{-1} (Graham et al. 1983; Shipley et al. 1992), and thus 10,000 kg of rye biomass ha^{-1} requires at least 90 to 110 kg of available N ha^{-1}. Cereal rye is recognized for its ability to proficiently scavenge residual inorganic soil N left in the soil from the previous season (Adeli et al. 2011). In regions with high soil organic matter content and cool, moist climates, residual N levels are likely sufficient to support optimum cereal rye biomass production. This is especially true of soils with a history of manure and legume cover crop use (Power and Doran 1984; Spargo et al. 2011). However, where residual soil inorganic N and/or mineralizable N levels are low, it may be necessary to fertilize cereal rye in order to produce biomass levels high enough to effectively suppress weeds.

In order to maximize N use efficiency and minimize residual N after cereal rye termination, it is important to target the period of highest N demand, which for cereals is between Zaddoks growth stage (GS)-30 and GS-60. Therefore, any supplemental N applied must be at or prior to GS-30 (stem elongation). This can be challenging with organically approved materials with low levels of soluble N. Many sources of N in organic systems also include P (e.g., dairy manure and poultry litter) that can constrain their use if nutrient management is a concern. The development of in-season indicators to predict cereal rye biomass potential, N demand, or both would better facilitate efficient N fertility management. Tiller density at GS-25 has been used as an effective predictor of fertilizer N need for wheat (Scharf and Alley 1993; Weisz et al. 2001) and may prove equally useful for cereal rye. In an ongoing experiment, we found that tiller density at GS-25 explained 58 and 66%, respectively, of the variability in rye biomass at GS-60 (Mirsky et al., unpublished data). Nitrogen management strategies that consider fertility source and temporal N demand must be developed in order to ensure efficient resource use and limit the reactive N load on the environment and subsequent weed populations.

CEREAL RYE GERMPLASM: TRAITS OF INTEREST

Cereal rye has not been explicitly bred to deliver any of the potential ecosystem services that are provided in this reduced-tillage organic system. This cereal has been bred for both forage use and as a grain crop. Key traits that would be desirable for use in our reduced-tillage organic system include a more moderate height or stronger stems that would lessen the amount of lodging (management constraints from lodging discussed below); greater allelochemical content (Reberg-Horton et al. 2005), which can also slow residue decomposition (Wagger et al. 1998); rapid early canopy development; and maturation times that match the desired planting dates for each region (Reberg-Horton et al. 2012). Breeding is more difficult with cereal rye than other small grains. As an obligate outcrosser, selfing is not possible and breeding approaches such as half-sib mating must be employed. Height and maturation time are highly heritable and easy to select for. Less is known about the heritability of allelochemical content, though substantial variation is believed to be present based on wheat (Wu et al. 2000); however, limited screening of rye germplasm has occurred (Reberg-Horton 2005). The demand for these desirable cover crop traits must increase to justify any future public or private breeding effort with cereal rye or other cover crops.

SOYBEAN PLANTING EQUIPMENT CHALLENGES

The high levels of cereal rye biomass required for persistent weed suppression can constrain soybean seed placement. Having equipment that can adequately plant soybean seed through high levels of cereal rye mulch is critical for soybean production success. In this cover crop–based system, soybean seed placement requires partial or full cutting of rye residue since the rolled cover crop can lay down in multiple angles when lodging occurs prior to rolling. As a result, planter and drill technology must be equipped with the appropriate coulters, planting unit gauge wheels, row cleaners (available on planters only), weight, and effective closing wheels to ensure accurate seed placement. In addition to working with the proper equipment for high-residue crop production, coordinating management decisions based on soil moisture levels may improve crop establishment and performance. The following subsections within this equipment section represent the evolution in planting technology used by the authors over the past 10 years. We summarize both unpublished data as well as information gathered through trial and error. The proper configuration of soybean planters and drills cannot be overstated because good soybean populations are necessary for not only maximizing crop yield but also minimizing weed competition.

Planters vs. Drills

There are trade-offs to using drills vs. planters for soybean planting into high levels of cereal rye biomass. Based on our experience with high cereal rye biomass levels, planters provide more consistent soybean populations compared to drills due to lower cutting resistance and greater precision with both seed metering and depth control. Even on drills with dual tool bars, there is less space between planting units, which limits options for residue management (i.e., row cleaners). The amount of possible down pressure per row is also less for drills due to the increased number of rows. However, reducing the time necessary for soybean canopy closure through narrow crop row spacing is an important cultural practice that aids in weed control (Hock et al. 2006). When possible, equipment selection should be based on cereal rye performance in a given year (Table 2). When rye biomass levels are moderate (4,000 to 6,000 kg ha^{-1}), a no-till drill can achieve good seed placement (Mirsky, unpublished data). As biomass levels increase, we find planters provide a better option. Configuring planters with split-row units to plant on a 40-cm row spacing may provide the best alternative (Reberg-Horton, personal communications).

Table 2: A decision framework for soybean row spacing, equipment type (drill or planter) and use of high-residue cultivation based on experimentation by authors

Cereal rye biomass levels		
Weed seed bank density	Low (3,000 to 6,000 kg ha^{-1})	High (7,000 to 12,000 kg ha^{-1})
Low	Drill	Splitter
High	Planter with cultivation	Splitter or planter with cultivation

Equipment Weight

Planting soybean into thick cereal rye mulch requires heavy-duty downforce springs or a pneumatic downforce system. With dry conditions, even greater downward force is required to penetrate through both the rye and soil to get adequate seed placement. Planters may need extra weight mounted on the frame. Smaller planters and drills (research plot or small farm scale; 3 m wide or less) have more difficulty with this problem due to their lower weight, while heavier equipment can be an advantage (Curran, personal communications). On heavier-textured soils (silt loam) in Pennsylvania and Maryland, 136 kg of weight is routinely added on each planter row unit using bags of sand or seed or steel weights in order to get sufficient soil penetration under drier conditions. Other planter modifications can include frame mounting one or two large (approximately 567 L) poly tanks that can be filled with water to increase planter weight.

Coulters

The type of coulters used on a drill or planter plays a critical role in ensuring good seed placement. On no-till planters and drills, the coulter functions to cut residue and prepare a better seedbed prior to the disk openers. For both drills and planters, several coulter types are available. In our work, we have relied primarily on lightly fluted to straight-edged coulters that are designed for residue cutting and soil penetration. We find wave or turbo-type coulters are less desirable due to their higher cutting surface area that often leads to hair-pinning of residue or ineffective cutting. Furthermore, such coulter types can result in greater soil disturbance that stimulates weed germination. Even, light fluting is a compromise between residue cutting ability and seedbed preparation. In high-residue environments, straight coulters or bubble coulters are more effective at cutting residue but cause more sidewall compaction (Reberg-Horton, personal communications). In our work, we have used no-till drills equipped with an integrally mounted coulter to cut residue and prepare a mini-seedbed.

Row Cleaners

Row cleaners can help improve stand establishment in heavy residue. In Pennsylvania, a 25% increase in soybean population was observed when using row cleaners in contrast to using the same planter without row cleaners (Mirsky, unpublished). Increased populations and faster emergence time resulted in either no difference in weed biomass or up to a 70% reduction (decrease by 700 kg ha^{-1}). That said, cover crop residue could wrap around the wheels of residue managers and reduce their usefulness. Although residue managers can improve crop stand, they also move the weed-suppressive cover crop mulch away from the crop row that is critical for weed control in organic no-till.

Closing Wheels

The type of closing wheel and mechanisms for downward pressure are particularly important in high-residue environments where it is more difficult to get adequate furrow closure and resulting seed-to-soil contact. Even with the heaviest of closing wheels, sealing the slit can be difficult at times and particularly under moist soil conditions. Poor furrow closure can not only reduce germination and emergence due to more extreme micro-climatic conditions, but also leave seeds and seedlings vulnerable to birds and other pests, which can severely reduce populations. Spading or spiked closing wheels rather than solid types can help close the seed slit in wet soils. In Beltsville and at Penn State, we have also used a single 38-cm spiked closing wheel alongside a smooth cast closing wheel on each row unit to help improve slit closure. Site-specific evaluations of closing wheels, down pressure configurations, and adjusting depth-gauge wheel settings, in different cover crop and soil moisture situations, can greatly improve soybean establishment and stands.

Management Timing and Crop–Soil Interactions

Beyond mechanical factors, soil and plant moisture greatly influences crop seed placement and establishment through high-residue mulches. In work by Nord et al. (2011), that included both a planter and a drill across five different soybean planting dates in the spring with increasing amounts of rye biomass, soybean stand differences were not linearly related to termination date as expected. Rather than being influenced solely by mulch biomass, soybean stand may have been more influenced by differences in rainfall patterns and subsequent soil moisture that, in turn, influenced planter or drill penetration, seed placement, and furrow closure. In additional work by Mischler et al. (2010), soybean populations were reduced in later terminated rye at one location presumably due to greater amounts of residue, while dry weather during an earlier planted treatment made it more difficult to achieve consistent soybean populations at another location. Such variability in soybean stand density between termination dates suggests that the technical challenges of planting into a thick layer of cover crop mulch are not trivial. We outline several important factors to consider.

The optimal timing of termination of cereal rye with a roller-crimper relative to planting soybean appears to be linked to regional factors. Delaying soybean planting by several days after rolling can allow the cereal rye to recover. The regrowth or recovering rye will often not grow parallel to rolling, thus requiring more extensive residue cutting with the planter or drill. Rolling and planting on the same day is attractive in regions with ample spring moisture. This approach allows the rolling and planting to occur simultaneously with the roller-crimper either mounted to the front of the tractor and the planter or drill to the rear, or both implements in the rear with the toolbars connected. Based on our experience, another advantage to a single roll/plant operation is that fresh plant material is easier to cut than aged surface residue. This is particularly true when aged surface mulches are wet. In drier locations or on sandier soils, rolling and allowing soil moisture to be replenished by precipitation is advised (Price et al. 2009). Several investigators recommend killing cover crops well before planting to allow precipitation to rebuild soil moisture (Hargrove and Frye 1987; Price et al. 2009; Reeves 2003). The delay in management can have ancillary benefits;Hammond and Cooper (1993) suggested no-till planting approximately 14 d after killing cereal rye to avoid seedcorn maggot damage in soybean. While our research team has not tested this practice, other authors have suggested planting first and rolling later because it does not require a high performance no-till planter (Bernstein et al. 2011). The authors find it generally more important to let soil moisture drive the timing of planting. Therefore whether roll/planting or roll-roll/plant, we time operations with a rain event to get maximum soybean establishment.

POTENTIAL FOR A SYNERGISTIC INTERACTION BETWEEN RYE RESIDUE AND SOYBEAN POPULATION

We tested the effects of combining two cultural management practices for improving weed suppression in organic soybean no-till planted into a rolled-crimped cereal rye cover crop (Ryan et al. 2011b). We used a factorial design to test for an interaction between the practices of increasing soybean planting rate and increasing cereal rye mulch, both of which can be accomplished by various means described above. Soybean was no-till planted at 0, 19, 37, 56, and 74 seeds m^{-2} and cereal rye residue was applied at five different rates representing 0, 0.5, 1, 1.5, and 2 times the normal rate (5,000 to 8,500 kg ha^{-1}). We hypothesized that the mulch would delay weed emergence, while allowing the soybean, with a larger seed size relative to weeds, to emerge first and develop a relatively more competitive leaf canopy. We observed a synergistic interaction between practices in two of the four site-years. This interaction occurred because increasing soybean density did not affect weed biomass in the absence of rye residue; however, when sufficient rye residue was included, increasing soybean density resulted in a decrease in weed biomass. Optimum weed suppression and yields were consistently observed when cereal rye biomass was 8,000 kg ha^{-1} and soybean was planted at the highest rate (740,000 seeds ha^{-1}). At very high mulch rates, soybean yield declined because of poor stands. This is an important example of how a synergistic interaction between two complementary cultural practices can be exploited for enhanced weed management in cover crop–based organic rotational no-till soybean production.

High Residue Cultivation

Despite the best cover crop management, high weed seed bank densities and perennial weed infestations constrain the success of the cover crop–based system. While crop rotation management is necessary for reducing weed seed banks and suppressing perennial weeds, high-residue cultivation is a viable option in cover crop–based organic rotational no-till. Results from several recent experiments illustrate the utility of high-residue cultivators for supplemental weed suppression in organic no-till planted soybean (Nord et al. 2011). High-residue cultivators are generally characterized by the presence of a coulter in front of the cultivator shank to cut through mulch, along with a relatively flat cultivator sweep that severs weed shoots from their roots just beneath the soil surface with minimum disturbance to the residue (Bowman 1997). Earlier research on inter-row high-residue cultivation was less successful because of excessive soil disturbance and antagonism between surface residue and cultivation efficacy (Teasdale and Rosecrance 2003). However, improvements to high-residue cultivators that minimize soil disturbance and better slice through surface residue have increased their utility. We conducted experiments in Beltsville, MD; Kutztown, PA; Raleigh, NC; and Rock Springs, PA to test the effects of using a high-residue cultivator for supplemental weed control in organic no-till planted soybean. Results showed that weed biomass can be five times higher in uncultivated soybean than when cultivated (Table 3). However, the effects of high-residue cultivation on yields have been mixed. High-residue cultivation increased yields, particularly when weed biomass was high. However, reductions in weed biomass did not consistently result in increased crop yields. High-residue cultivators do not always kill weeds, but instead slow their growth and subsequent impact on yield. Furthermore, untimely high-residue cultivation can kill weeds after the competitive effect on the crop has already been realized. The result would be differences in weed biomass but less of an impact on yield. Future work should investigate the effects timing of high-residue cultivation has on weed–crop competition and when it is economically beneficial to cultivate.

Table 3: Cereal rye biomass, weed biomass, density, or cover and crop yield improvement across four sites on the eastern United States including Pennsylvania State University (Rock Springs, PA); USDA-ARS (Beltsville, MD); The Rodale Institute (Kutztown, PA); and North Carolina State University (Raleigh, NC). Values within parentheses are standard errors of the mean

Site	**Years**	**Cereal rye biomass**	**Weed biomass, density, or cover**		**Crop yield improvement**
			Uncultivated	Cultivated	
			kg ha^{-1}		
Beltsville, MD	2009	7,012 (376)	1,952 (465)	674 (260)	39
Rock Springs, PA	2009	7,722 (451)	1,601 (234)	974 (223)	NS
	2010	7,548 (483)	1,883 (280)	859 (144)	NS
	2011	6,719 (225)	30 (25)	41 (5)	NS
	2011	1,674 (650)	1,933 (211)	1545 (9)	54
Kutztown, PA	2010	10,608 (630)	893 (225)	139 (70)	23
	2011	5,974 (473)	2,887 (563)	1,525 (303)	18
			% *cover* or weed (plants m^{-2})		%
			density		
Raleigh, NC	2008	6,606	61%	44%	NS
	2009	8,066	1.2	0.99	NS
	2009	8,366	0.13	0.006	NS

Integrating No-Till Soybean into Existing Rotations

The research described in this paper defines a potentially viable rotational no-till system for organic soybean production that has proven effective in short-term experiments. Challenges still remain in determining the best fit for this system in long-term field crop rotations. Typical field crop rotations could include corn–soybean, corn–soybean–wheat, and corn–soybean–wheat–3 yr alfalfa (*Medicago sativa* L.). Therefore, cropping system integration of rotational no-till soybean is constrained by several factors, including timing of previous cash crop harvest (i.e., corn) to ensure timely cereal rye establishment and timely soybean planting for successful subsequent small grain or cover crop establishment. As a result, evaluation of the rotational no-till soybean system within a long-term crop rotation is necessary to confirm optimal, regionally specific crop and cover crop management timing.

Integration of rotational no-till soybean production into organic grain rotations also requires increasing soil N mineralization levels for adequate cereal rye biomass production. This is increasingly important when crop production is on coarser textured soils or those with low soil organic matter levels. Nitrogen availability and its management result in trade-offs between producing optimal weed suppressive mulches and minimizing weed performance. As stated previously, low N availability will impact weeds, but to a greater extent the cover crop biomass required for weed suppression. Insufficient N for adequate rye biomass production may be particularly problematic during the transition phase to organic when soil organic matter may be relatively low. During this stage, supplemental N may be warranted to ensure adequate rye growth. However, organic farms can build soil organic levels significantly over time, meaning more N could be available for rye cover crops on farms that have been organic longer. Building up soil mineralization potential is important because most growers would not consider fertilizing their cover crops. However, higher N mineralization from soil may also impact the weed-suppressive ability of the cereal rye mulch by overcoming the duration of N immobilization from the surface mulch. Furthermore, the magnitude of these effects is going to depend on soil and crop management history (Mallory and Friffin 2007).

Lastly, using a multitactic approach to weed management will be critical to the success of cover crop–based organic rotational no-till soybean production. As part of a holistic systems approach, mechanical tactics should be integrated with cultural weed management practices. The identification of synergistic interactions between cultural practices, as discussed for rye residue and soybean population, is particularly important for designing effective weed management for organic agriculture. Embedding this no-till organic soybean system within a suitable crop and tillage rotation is particularly important to realizing the potential benefits of this system. Use of a phenologically diverse crop rotation has been shown to be particularly important for reducing the weed seed bank (Teasdale et al. 2004) and increasing crop yields (Cavigelli et al. 2008) in organic grain production. Multitactic weed management would be most effective when weed seed bank densities are low, but would be critical when densities are high, requiring sequences of cultural and mechanical practices to be implemented in combination to drive down weed seed banks. Equally important would be rotational use of tillage during selected cropping sequences to disrupt weed species favored by absence of tillage (e.g., perennials) as well as permit incorporation of organic amendments and facilitate planting smaller-seeded crops such as forages.

REFERENCES

1. Adeli, A., H. Tewolde, J. N. Jenkins, and D. E. Rowe. 2011. Cover crop use for managing broiler litter applied in the fall. Agron. J. 103:200–210.
2. Anser, G. P. and A. R. Townsend. 1997. The decoupling of terrestrial carbon and nitrogen cycles. BioScience 47:226–234.
3. Ashford, D. L. and D. W. Reeves. 2003. Use of a mechanical roller crimper as an alternative kill method for cover crop. Am. J. Alt. Agric. 18:37–45.
4. Barnes, J. P. and A. R. Putnam. 1983. Rye residues contribute weed suppression in no-tillage cropping systems. J. Chem. Ecol. 9:1045–1057.

5. Barnes, J. P. and A. R. Putnam. 1987. Role of benzoxazinones in allelopathy by rye. J. Chem. Ecol. 13:889–906.
6. Bernstein, E. R., J. L. Posner, D. E. Stoltenberg, and J. L. Hedtcke.2011. Organically managed no-tillage rye-soybean systems: agronomic, economic, and environmental assessment. Agron. J. 103:1169–1179.
7. Blackshaw, R. E., L. J. Molnar, and H. H. Janzen. 2004. Nitrogen fertilizer timing and application method affect weed growth and competition with spring wheat. Weed Sci. 52:614–622. BioOne
8. Bowman, G. 1997. Steel in the Field: A Farmer's Guide to Weed-Management Tools. Beltsville, MD: Sustainable Agriculture Network.
9. Boyd, N. S., E. B. Brennan, R. F. Smith, and R. Yokota. 2009. Effect of seeding rate and planting arrangement on rye cover crop and weed growth. Agron. J. 101:47–51.
10. Burgos, N. R., R. E. Talbert, and J. D. Mattice. 1999. Cultivar and age differences in the production of allelochemicals by *Secale cereale*. Weed Sci. 47:481–485.
11. Clark, A. 2007. Managing cover crop profitably. 3rd edition. Beltsville, MD: Sustainable Agriculture Network.
12. Creamer, N. G. and S. M. Dabney. 2002. Killing cover crops mechanically: review of recent literature and assessment of new research results. Am. J. Alt. Agric. 17:32–40.
13. Cavigelli, M. A., S. B. Mirsky, J. R. Teasdale, J. T. Spargo, and J. Doran. 2013. Organic management systems to enhance ecosystem services. Renew. Agric. Food Syst. *In press*.
14. Cavigelli, M. A., J. R. Teasdale, and A. E. Conklin. 2008. Long-term agronomic performance of organic and conventional field crops in the mid-Atlantic region. Agron. J. 100:785–794.
15. Dabney, S. M., J. A. Delgado, J. J. Meisinger, H. H. Schomberg, M. A. Liebig, T. Kaspar, J. Mitchell, and W. Reeves. 2010. Using cover crops and cropping systems for nitrogen management. Pages 231–282 *in* J. A. Delgado and R. F. Follett, eds. Advances in Nitrogen Management for Water Quality. Ankeny, IA: SWCS.
16. Davis, A. 2010. Cover crop roller-crimper contributes to weed management in no-till soybean. Weed Sci. 58:300–309. BioOne

17. Decourtye, A., E. Mader, and N. Desneux. 2010. Landscape enhancement of floral resources for honey bees in agro-ecosystems. Apidologie 41:264–277.
18. Delate, K., D. Cwach, and C. Chase. 2012. Organic no-till system effects on organic soybean, corn, and tomato production and economic performance in Iowa. Renew Agric. Food Syst. 27(special issue 01):49–59.
19. Derpsch, R., C. H. Roth, N. Sidiras, and U. Köpke. 1991. Controle da erosão no Paraná, Brazil: Sistemas de cobertura do solo, plantio directo e prepare conservacionista do solo. Eschborn, Germany:Deutsche Gesellschaft für Technische Zusammenarbeit (GTZ) GmbH.
20. Facelli, J. M. and S.T.A. Pickett. 1991. Plant litter: light interception and effects on an old-field plant community. Ecology 72:1024–1031.
21. Franzluebbers, A. J. 2002. Water infiltration and soil structure related to organic matter and its stratification with depth. Soil Tillage Res. 66:197–205.
22. Franzluebbers, A. J. 2005. Soil organic carbon sequestration and agricultural greenhouse gas emissions in the southeastern USA. Soil Tillage Res. 83:120–147.
23. Frey, S. D., E. T. Elliott, K. Paustian, and G. A. Peterson. 2000. Fungal translocation as a mechanism for soil nitrogen inputs to surface residue decomposition in a no-tillage agroecosystem. Soil Biol. Biochem. 32:689–698.
24. Graham, R., P. Geytenbeek, and B. Radcliffe. 1983. Responses of triticale, wheat, rye and barley to nitrogen fertilizer. Aust. J. Exp. Agric. 23:73–79.
25. Hammond, R. B. and R. L. Cooper. 1993. Interaction of planting times following the incorporation of a living, green cover crop and control measures on seedcorn maggot populations in soybean. Crop Prot. 12:539–543.
26. Hargrove, W. L. and W. W. Frye. 1987. The need for legume cover crops in conservation tillage production. Pages 1–5 *in* J. F. Power, ed. The Role of Legumes in Conservation Tillage Systems. Ankeny, IA: Soil Conservation Society of America.

27. Henson, J. F. and L. S. Jordan. 1982. Wild oat (*Avena fatua*) competition with wheat (*Triticum aestivum* and *T. turgidum durum*) for nitrate. Weed Sci. 30:297–300.

28. Hock, S. M., J. L. Lindquist, A. R. Martin, and S. Z. Knezevic. 2006. Soybean row spacing and weed emergence time influence weed competitiveness and competitive indices. Weed Sci. 54:38–46.

29. Kornecki, T. S., A. J. Price, and R. L. Raper. 2006. Performance of different roller designs in terminating rye cover crop and reducing vibration. Appl. Eng.. Agric. 22:633–641.

30. Kornecki, T. S., A. J. Price, R. L. Raper, and F. J. Arriaga. 2009. New roller crimper concepts for mechanical termination of cover crops. Renew. Agric. Food Syst. 24:165–173.

31. Kornecki, T. S., R. L. Raper, F. J. Arriaga, E. B. Schwab, and J. S. Bergtold. 2009. Impact of rye rolling direction and different no-till row cleaners on cotton emergence and yield. Trans. ASABE 52:383–391.

32. Macías, F. A., D. Marín, A. Oliveros-Bastidas, D. Castellano, A. M. Simonet, and J. M. G. Molinillo. 2005. Structure-activity relationship studies of benzoxazinones, their degradation products, and analogues. Phytotoxicity on standard target species. J. Agric. Food Chem. 53:538–548.

33. Mallory, E.B. and T.S. Griffin. 2007. Impacts of soil amendment history on nitrogen availability from manure and fertilizer. Soil Sci. Soc. Am. J. 71:964–973.

34. Mirsky, S. B., W. S. Curran, D. A. Mortensen, M. R. Ryan, and D. L. Shumway. 2009. Control of cereal rye with a roller/crimper as influenced by cover crop phenology. Agron. J. 101:1589–1596.

35. Mirsky, S. B., W. S. Curran, D. A. Mortensen, M. R. Ryan, and D. L. Shumway. 2011. Timing of cover crop management effects on weed suppression in no-till planted soybean using a roller-crimper. Weed Sci. 59:380–389. BioOne

36. Mirsky, S. B., M. R. Ryan, W. S. Curran, J. R. Teasdale, J. Maul, J. T. Spargo, J. Moyer, A. M. Grantham, D. Weber, T. R. Way, and G. G. Camargo. 2012. Conservation tillage issues: cover crop-based organic rotational no-till grain production in the mid-Atlantic region, USA. Renew. Agric. Food Syst 27:31–40.

37. Mischler, R. A., W. S. Curran, S. W. Duiker, and J. A. Hyde. 2010. Use of a rolled-rye cover crop for weed suppression in no-till soybeans. Weed Technol. 24:253–261. BioOne

38. Mohler, C. L. 1996. Ecological basis for the cultural control of annual weeds. J. Prod. Agric. 9:468–474.

39. Mohler, C. L. and J. R. Teasdale. 1993. Response of weed emergence to rate of *Vicia villosa* Roth and *Secale cereale* L. residue. Weed Res. 33:487–499.

40. Moore, M. J., T. J. Gillespie, and C. J. Swanton. 1994. Effect of cover crop mulches on weed emergence, weed biomass, and soybean (*Glycine max*) development. Weed Technol. 8:512–518.

41. [NASS] National Agriculture Statistics Service. 2012.http://www.nass.usda.gov/Data_and_Statistics/Quick_Stats/index.asp. Accessed June 3, 2012.

42. Nord, E. A., W. S. Curran, D. A. Mortensen, S. B. Mirsky, and B. P. Jones. 2011. Integrating multiple tactics for managing weeds in high residue no-till soybean. Agron. J. 103:1542–1551.

43. Nord, E. A., M. R. Ryan, W. S. Curran, D. A. Mortensen, and S. B. Mirsky. 2012. Weed emergence periodicity mediates interaction between management system and planting date in no-till planted soybean. Weed Sci. 60:624–633.

44. Parsch, L. D., T. C. Keisling, P. A. Sauer, L. R. Oliver, and N. S. Crabtree. 2001. Economic analysis of conservation and conventional tillage cropping systems on clayey soil in eastern Arkansas. Agron. J. 93:1296–1304.

45. Pesant, A. R., J. L. Dionne, and J. Genest. 1987. Soil and nutrient losses in surface runoff from conventional and no-till corn systems. Can. J. Soil Sci. 67:835–843.

46. Phelan, L., D. Stinner, C. Nacci, and D. McCartney. 2008. Application of the niche concept to organic weed management. Pages 29–30 *in*Proceedings of the Midwest Organic Research Symposium. La Crosse, WI: Organic Farming Research Foundation.

47. Power, J. F. and J. W. Doran. 1984. Nitrogen use in organic farming. Pages 585–568 *in* R. D. Hauck, ed. Nitrogen in Crop Production.Madison, WI: ASA, CSSA, and SSSA.

48. Price, A. J., F. J. Arriaga, R. L. Raper, K. S. Balkcom, T. S. Kornecki, and D. W. Reeves. 2009. Comparison of mechanical and chemical winter cereal cover crop termination systems and cotton yield in conservation agriculture. Cotton Sci. 13:238–245.

49. Price, A. J., K. S. Balkcom, and S. A. Culpepper. 2011. Glyphosate-resistant Palmer amaranth: a threat to conservation tillage. J. Soil .Water Conserv. 66:265–275.

50. Qi, Z. and M. J. Helmers. 2009. Soil water dynamics under winter rye cover crop in central Iowa. Vadose Zone J. 9:53–60.

51. Reberg-Horton, S. C., J. D. Burton, D. A. Danehower, G. Ma, D. W. Monks, J. P. Murphy, N. N. Ranells, J. D. Williamson, and N. G. Creamer. 2005. Changes over time in the allelochemical content of ten cultivars of rye. J. Chem. Ecol. 31:179–193.

52. Reberg-Horton, S. C., J. Grossman, T. S. Kornecki, A. D. Meijer, A. J. Price, G. T. Place, and T. M. Webster. 2012. Utilizing cover crop mulches to reduce tillage in organic systems in the Southeast. Renew. Agric. Food Syst. 27:41–48.

53. Reeves, D. W. 2003. A Brazilian model for no-tillage cotton production adapted to the southeastern USA. Pages 372–374 *in* Proceedings of the Second World Congress on Conservation Agriculture. Athens,Georgia.

54. Rice, C. P., G. Cai, and J. R. Teasdale. 2012. Fate of benzoxazinoids in soil treated with rye cover crop. J. Agric. Food Chem. 60:4471–4479.

55. Ryan, M. R. 2010. Energy Usage, Greenhouse Gases, and Multi-Tactical Weed Management in Organic Rotational No-Till Cropping Systems. Ph.D. dissertation. University Park, PA: The Pennsylvania State University.

56. Ryan, M. R., W. S. Curran, A. M. Grantham, L. K. Hunsberger, S. B. Mirsky, D. A. Mortensen, E. A. Nord, and D. O. Wilson. 2011a. Effects of seeding rate and poultry litter on weed suppression from a rolled cereal rye cover crop. Weed Sci. 59:438–444. BioOne

57. Ryan, M. R., S. B. Mirsky, D. A. Mortensen, J. R. Teasdale, and W. S. Curran. 2011b. Potential synergistic effects of cereal rye biomass and soybean planting density on weed suppression. Weed Sci. 59:238–246. BioOne

58. Ruffo, M. L. and G. A. Bollero. 2003. Modeling rye and hairy vetch residue decomposition as a function of degree days and decomposition days. Agron. J. 95: 900–997.

59. Scharf, P. C. and M. M. Alley. 1993. Spring nitrogen on winter-wheat. II. A flexible multicomponent rate recommendation system. Agron. J. 85:1186–1192.

60. Shipley, P. R., J. J. Messinger, and A. M. Decker. 1992. Conserving residual corn fertilizer nitrogen with winter cover crops. Agron. J. 84:869–876.

61. Snapp, S. S., S. M. Swinton, R. Labarta, D. Mutch, J. R. Black, R. Leep, J. Nyiraneza, and K. O'Neil. 2005. Evaluating cover crops for benefits, costs and performance within cropping system niches. Agron. J. 97:322–332.

62. Spargo, J. S., M. M. Alley, R. F. Follett, and J. V. Wallace. 2008. Soil carbon sequestration with continuous no-till management of grain cropping systems in the Virginia Coastal Plain. Soil Tillage Res. 100:133–144.

63. Spargo, J. T., M. A. Cavigelli, S. B. Mirsky, J. E. Maul, and J. J. Meisinger. 2011. Mineralizable soil nitrogen and labile soil organic matter in diverse long-term cropping systems. Nutr. Cycl. Agroecosyst. 90:253–266.

64. Teasdale, J. R., C. B. Coffman, and R. W. Mangum. 2007. Potential long-term benefits of no-tillage and organic cropping systems for grain production and soil improvement. Agron. J. 99:1297–1305.

65. Teasdale, J. R., R. W. Mangum, J. Radhakrishnan, and M. A. Cavigelli.2004. Weed seedbank dynamics in three organic farming crop rotations. Agron. J. 96:1429–1435.

66. Teasdale, J. R. and C. L. Mohler. 1993. Light transmittance, soil–temperature, and soil–moisture under residue of hairy vetch and rye. Agron. J. 85:673–680.

67. Teasdale, J. R. and C. L. Mohler. 2000. The quantitative relationship between weed emergence and the physical properties of mulches. Weed Sci. 48:385–392. BioOne

68. Teasdale, J. R., C. P. Rice, G. Cai, and R. W. Magnum. 2012. Expression of allelopathy in the soil environment: soil concentration and activity of benzoxazinoid compounds

released by rye cover crop residue. J. Plant Ecol. http://dx.doi.org/10.1007/s11258-012-0057-x.

69. Teasdale, J. R. and R. C. Rosecrance. 2003. Mechanical versus herbicidal strategies for killing a hairy vetch cover crop and controlling weeds in minimum-tillage corn production. Am. J. Alt. Agric. 18:95–102.

70. Tungate, K. D., M. G. Burton, D. J. Susko, S. M. Sermons, and T. W. Rufty. 2006. Altered weed reproduction and maternal effects under low-nitrogen fertility. Weed Sci. 54:847–853. BioOne

71. U.S. Environmental Protection Agency. 2007. Mid-Atlantic Water: Basic Information about Agriculture.http://www.epa.gov/reg3wapd/Agriculture/basicinfoaboutag.html. Accessed June 22, 2011.

72. Wagger, M. G., M. L. Cabrera, and N. N. Ranells. 1998. Nitrogen and carbon cycling in relation to cover crop residue quality. J. Soil Water Conserv. 53:214–218.

73. Ward, M. J., M. R. Ryan, W. S. Curran, M. E. Barbercheck, and D. A. Mortensen. 2011. Cover crops and disturbance influence activity-density of *Amara aenea* and *Harpalus pensylvanicus* (Coleoptera: Carabidae). Weed Sci. 59:76–81. BioOne

74. Weisz, R., C. R. Crozier, and R. W. Heiniger. 2001. Optimizing nitrogen application timing in no-till soft red winter wheat. Agron. J. 93:435–442.

75. Wells, M. S., S. C. Reberg-Horton, and A. N. Smith. 2010. Nitrogen immobilization in a rye (*Secale cereale* L.) roll-killed system. Pages106–114 *in* Proceedings of the International Annual Meetings of the ASA-CSSA-SSSA. Madison. WI: ASA-CSSA-SSSA.

76. Whitehouse, S. E., A. DiTommaso, L. E. Drinkwater, and C. L. Mohler.2009. Changes in weed vigor and growth in response to carbon and nitrogen ratio manipulation. Page 47 *in* Proceedings of the Sixty-Third Annual Meeting of the Northeastern Weed Science Society.Fredericksburg, PA: Northeastern Weed Science Society.

77. Winsten, J. R., C. D. Kerchner, A. Richardson, A. Lichau, and J. M. Hyman. 2010. Trends in the Northeast dairy industry: large-scale modern confinement feeding and management-intensive grazing. J. Dairy Sci. 93:1759–1769.

78. Wu, H., J. Pratley, D. Lemerle, and T. Haig, 2000. Evaluation of seedling allelopathy in 453 wheat (*Triticum aestivum*) accessions against annual rye grass (*Lolium rigidum*) by the equal-compartment- agar method. Austral. J. Agric. Res. 51:937–944

Citations

CHAPTER 1

F. Baret, V. Houles, and M. Guerif, Quantification of Plant Stress using Remote Sensing Observations and Crop Models: The Case of Nitrogen Management, J. Exp. Bot. (2007) 58 (4): 869-880 first published online January 13, 2007 doi:10.1093/jxb/erl231.

CHAPTER 2

Xiao-Tang Ju, Guang-Xi Xing, Xin-Ping Chen, Shao-Lin Zhang, Li-Juan Zhang, Xue-Jun Liu, Zhen-Ling Cui, Bin Yin, Peter Christie, Zhao-Liang Zhu, and Fu-Suo Zhang. Reducing Environmental Risk by Improving N Management in Intensive Chinese Agricultural Systems PNAS 2009 106 (9) 3041-3046; published ahead of print February 17, 2009, doi:10.1073/pnas.0813417106.

CHAPTER 3

John King, Alan Gay, Roger Sylvester Bradley, Ian Bingham, John Foulkes, Peter Gregory, and David Robinson, Modelling Cereal Root Systems for Water and Nitrogen Capture: Towards an Economic Optimum Ann Bot (2003) 91 (3): 383-390 first published online January 8, 2003 doi:10.1093/aob/mcg033.

CHAPTER 4

Matt A. Sanderson and Paul R. Adler, Perennial Forages as Second Generation Bioenergy Crops, Matt A. Sanderson and Paul R. Adler, Int. J. Mol. Sci. 2008, 9, 768-788; DOI: 10.3390/ijms9050768.

CHAPTER 5

Gary P. Munkvold, Richard L. Hellmich, Larry G. Rice, Comparison of Fumonisin Concentrations in Kernels of Transgenic Bt Maize Hybrids and Nontransgenic Hybrids, Plant Disease / Vol. 83 No. 2 Publication no. D-1998-1130-02R

CHAPTER 6

Jianchang Yang and Jianhua Zhang, Crop Management Techniques to Enhance Harvest Index in Rice J. Exp. Bot. first published online April 25, 2010 doi:10.1093/jxb/erq112.

CHAPTER 7

C. N. Hewitt, A. R. MacKenzie, P. Di Carlo, C. F. Di Marco, J. R. Dorsey, M. Evans, D. Fowler, M. W. Gallagher, J. R. Hopkins, C. E. Jones, B. Langford, J. D. Lee, A. C. Lewis, S. F. Lim, J. McQuaid, P. Misztal, S. J. Moller, P. S. Monks, E. Nemitz, D. E. Oram, S. M. Owen, G. J. Phillips, T. A. M. Pugh, J. A. Pyle, C. E. Reeves, J. Ryder, J. Siong, U. Skiba, and

D. J. Stewart, Nitrogen Management is Essential to Prevent Tropical Oil Palm Plantations from Causing Ground-level Ozone Pollution, PNAS 2009 106 (44) 18447-18451; published ahead of print October 19, 2009, doi:10.1073/pnas.0907541106.

CHAPTER 8

S. M. Shirazi, Zulkifli Yusop, N. H. Zardari, and Z. Ismail, Effect of Irrigation Regimes and Nitrogen Levels on the Growth and Yield of Wheat, Advances in Agriculture, vol. 2014, Article ID 250874, 6 pages, 2014. doi:10.1155/2014/250874.

CHAPTER 9

Fabrício Pinheiro Povh and Wagner de Paula Gusmão dos Anjos (2014). Optical Sensors Applied in Agricultural Crops, Optical Sensors - New Developments and Practical Applications, Dr Moh. Yasin (Ed.), ISBN: 978-953-51-1233-4, InTech, DOI: 10.5772/57145.

CHAPTER 10

Corjan Brink, Ekko van Ierland, Leen Hordijk, and Carolien Kroeze, Cost-Effective Emission Abatement in Europe Considering Interrelations in Agriculture, TheScientificWorldJOURNAL, vol. 1, pp. 814-821, 2001. doi:10.1100/tsw.2001.295.

CHAPTER 11

Xiaoyuan Yan, Chaopu Ti, Peter Vitousek, Deli Chen, Adrian Leip, Zucong Cai, and Zhaoliang Zhu, Fertilizer Nitrogen Recovery Efficiencies in Crop Production Systems of China with and without Consideration of the Residual Effect of Nitrogen, doi:10.1088/1748-9326/9/9/095002.

CHAPTER 12

Steven B. Mirsky, Matthew R. Ryan, John R. Teasdale, William S. Curran, Chris S. Reberg-Horton, John T. Spargo, M. Scott Wells, Clair L. Keene, and Jeff W. Moyer, Overcoming Weed Management Challenges in Cover Crop–Based Organic Rotational No-Till Soybean Production in the Eastern United States, doi: http://dx.doi.org/10.1614/WT-D-12-00078.1.

Index